Modelling Rock Fracturing Processes

Modelling Rock Fracturing Processes

Editor

Ali Asgar Samara

Modelling Rock Fracturing Processes

Edited by **Ali Asgar Samara**

Printed in 2017

ISBN: 978-1-68117-164-7
Library of Congress Control Number: 2015951144

© 2016 by
SCITUS Academics LLC,
616, Corporate Way, Suite 2, 4766,
Valley Cottage, NY 10989

www.scitusacademics.com

Notice

Preface

Hydraulic fracturing is a well-stimulation technique in which rock is fractured by a pressurized liquid. The process involves the high-pressure injection of 'fracking fluid' (primarily water, containing sand and other proppants suspended with the aid of gelling agents) into a wellbore to create cracks in the deep-rock formations through which natural gas, petroleum, and brine will flow more freely.

This book describes a unique approach using the principles of rock fracture mechanics to investigate the behaviour of fractured rock masses for rock engineering purposes. Rock fracture mechanics, a promising outgrowth of rock mechanics and fracture mechanics, has developed rapidly in recent years, driven by the need for in-depth understanding of rock mass failure processes in both fundamental research and rock engineering designs. Today, as rock engineering extends into many more challenging fields (like mining at depth, radioactive waste disposal, geothermal energy, and deep and large underground spaces), it requires knowledge of rock masses, complex coupled thermal hydraulic chemical mechanical processes. Rock fracture mechanics play a crucial role in these complex coupled processes simply because rock fractures are the principal carrier and common interface. To date, the demand for rock fracture mechanics based design tools has outstripped the very limited number of numerical tools available.

The book also presents the fundamentals of thermo-mechanical coupling and hydro-mechanical coupling. Formulations of multiple regional mechanical, thermal and hydraulic functions, which allow analyses of fracture mechanics problems for structures made of brittle, rock-like materials, are provided. In addition, instructive examples of code verification and applications are presented.

Table of Contents

Chapter 1

Fracture Development around Deep Underground Excavations: Insights from FDEM Modelling

Andrea Lisjak[1], Daniel Figi[2], Giovanni Grasselli[2]

[1] Geomechanica Inc., 90 Adelaide Street West, Suite 300, M5H3V9 Toronto, ON, Canada

[2] Department of Civil Engineering, University of Toronto, 35 St. George Street, M5S1A4 Toronto, ON, Canada

ABSTRACT

Over the past twenty years, there has been a growing interest in the development of numerical models that can realistically capture the progressive failure of rock masses. In particular, the investigation of damage development around underground excavations represents a key issue in several rock engineering applications, including tunnelling, mining, drilling, hydroelectric power generation, and the deep geological disposal of nuclear waste. The goal of this paper is to show the effectiveness of a hybrid finite-discrete element method (FDEM) code to simulate the fracturing mechanisms associated with the excavation of underground openings in brittle rock formations. A brief review of the current state-of-the-art modelling approaches is initially provided, including the description of selecting continuum- and discontinuum-based techniques. Then, the influence of a number of factors, including mechanical and in situ stress anisotropy, as well as excavation geometry, on the simulated damage is analysed for three different geomechanical scenarios. Firstly, the fracture nucleation and growth process under isotropic rock mass conditions is simulated for a circular shaft. Secondly, the influence of mechanical anisotropy on the development of an excavation damaged zone (EDZ) around a tunnel excavated in a layered rock formation is considered. Finally, the interaction mechanisms between two large caverns of an underground hydroelectric power station are investigated, with particular emphasis on the rock mass response sensitivity to the pillar width and excavation sequence. Overall, the numerical results indicate that FDEM simulations can provide unique geomechanical insights in cases where an explicit consideration of fracture and fragmentation processes is of paramount importance.

INTRODUCTION

The stability of deep underground excavations is a common issue in a variety of rock engineering fields, including mining, tunnelling, hydroelectric power generation, and nuclear waste disposal. Furthermore, the deformation and failure of underground openings, such as boreholes, are of great importance in the drilling industry associated with hydrocarbon extraction and geothermal production. In tunnelling and mining operations, the stability of underground openings directly affects the choice of the excavation method and sequence, as well as the design of support and reinforcement measures. In the case of underground hydroelectric power stations, the rock mass behaviour is strongly affected by complex interaction mechanisms between multiple caverns. In the context of the deep geological disposal of nuclear waste, one main concern is that the disturbed zone around the excavations, namely the excavation damaged zone (EDZ), may negatively impact the hydro-mechanical behaviour of the rock mass, thus affecting its isolation properties and, as a consequence, the long-term safety of the repository.

Analytical solutions can be used to determine the stress and deformation fields around underground excavations (Brady and Brown, 2006). However, closed-form solutions are available only for simple excavation shapes (e.g. circular, elliptical) and under highly simplifying mechanical assumptions, such as perfect elasticity and homogeneity. Therefore, in engineering practice, numerical models are frequently used to analyse and predict the rock mass behaviour. In computational geomechanics, the numerical approaches are commonly classified as (i) continuum methods and (ii) discontinuum (or discrete) methods (Jing and Hudson, 2002). Conventionally, numerical models based on continuum mechanics are employed to simulate rock mass response to excavation process (e.g. Mizukoshi and Mimaki, 1985, Eberhardt, 2001 and Cai and Kaiser, 2014). However, their ability to consider the rock mass discontinuities remains somewhat limited. Although joint elements can be integrated into the continuum formulation (Hammah et al., 2008), typically only small deformations can be correctly captured due to the lack of contact detection and interaction algorithms (Cundall and Hart, 1992). On the other hand, discrete models may provide a more realistic representation of the physical behaviour observed in the field and, specifically, of the intrinsically discontinuous nature of the rock mass (Barton, 2011). Among the available discrete numerical approaches, the hybrid finite-discrete element method (FDEM) (Munjiza, 2004 and Mahabadi, 2012) captures material failure by explicitly considering fracture nucleation and propagation, as well as the interaction of pre-existing and newly-created discrete rock blocks.

In this study, FDEM simulations are used to obtain unique insights into the failure process around deep underground excavations for three different geomechanical scenarios. Firstly, the stress-driven fracturing process of a circular shaft excavated in a homogeneous and isotropic rock is analysed. Secondly, the influence of mechanical anisotropy on the development of an EDZ around a tunnel in a layered rock formation is considered. Thirdly, the interaction mechanisms between two adjacent underground caverns are investigated, with particular emphasis on the rock mass response sensitivity to the in situ stress anisotropy, pillar width and excavation sequence.

REVIEW OF AVAILABLE MODELLING APPROACHES

Numerical modelling in rock engineering is a challenging task owing to several characteristics of the rock mass behaviour. Firstly, the stress-strain response of the rock material under uniaxial compression is highly non-linear (Martin, 1997 and Jaeger et al., 2007). The initial strain hardening associated with the closure of voids and pre-existing microcracks is typically followed by a nearly linear stress-strain portion. Subsequently, the nucleation, propagation, and coalescence of microcracks lead to the loss of linearity, strain localization and the formation of macroscopic fractures. Upon reaching the peak strength, strain softening is associated with brittle rupture phenomena.

Secondly, the rock failure process is significantly influenced by confining pressure. Under unconfined compression (brittle) failure tends to occur in the form of axial splitting, while, under increasing confinement, the rock exhibits a more ductile behaviour accompanied by shear band formation.

Thirdly, the failure process observed at the laboratory-scale is further complicated at the rock mass level, where the behaviour is often influenced by the presence of discontinuities, such as joints, fractures, bedding planes, and tectonic structures. Discontinuities represent mechanical weaknesses of the rock mass and hence have a crucial effect on its deformability, strength, failure, and permeability properties (Hudson and Harrison, 1997). Moreover, the presence of discontinuities may add kinematic constraint on the deformation and failure mode of rock mass structures (Hoek et al., 1995).

Continuum approaches
The most commonly adopted numerical methods are the continuum-based approaches, such as the finite difference method (FDM), the finite element method (FEM), and the boundary element method (BEM). While FDM uses the differential form of the governing partial differential equations, FEM and BEM are based on their integral form and require solving a global equation system (Peiro and Sherwin, 2005). Continuum methods are suitable tools

for simulating the stress and deformation fields around underground excavations. However, due to the lack of an internal length scale, standard strength-based, strain-softening constitutive relationships cannot reproduce the localisation of failure, as the underlying mathematical problem becomes ill-conditioned (de Borst et al., 1993).

To overcome the above limitations, different enrichment approaches, such as higher-order constitutive laws (e.g. Masin, 2005), Cosserat micro-polar models (e.g. Mühlhaus and Vardoulakis, 1987), non-local models (e.g. Bažant and Pijaudier-Cabot, 1988), and meshfree methods (e.g. Rabczuk and Belytschko, 2004, Rabczuk and Belytschko, 2007,Zhuang et al., 2012 and Zhuang et al., 2014) have been introduced. Recently, techniques such as the generalised finite element method (GFEM) (e.g. Strouboulis et al., 2000) and the extended finite element method (XFEM) (e.g. Möes and Belytschko, 2002), based on addition of non-polynomial shape functions to the classical FEM formulation, have been adopted for rock mechanics applications. Belytschko et al. (2001)used XFEM to investigate the stability of a tunnel in a jointed rock mass by modelling the fractures as interior displacement discontinuities. A similar approach was adopted byDeb and Das (2010) to numerically analyse a circular tunnel intersected by a joint plane. XFEM has also been successfully employed to simulate the propagation of cohesive cracks within continuum finite element models (Möes and Belytschko, 2002 and Zhang and Feng, 2011). Unlike conventional fracture-mechanics-based studies (e.g. Steer et al., 2011), in XFEM the discontinuities are completely independent of the finite element mesh and, therefore, remeshing is not required. However, the technique is, in general, not well suited to capture the interaction of multiple, arbitrarily located discontinuities, as well as large-scale material flow and motion (Karekal et al., 2011).

Another class of continuum-based approaches is represented by damage mechanics models, which capture the heterogeneous nature of rocks by statistically distributing defects into numerical domain. Several variations of this technique have been implemented in FEM (Tang and Kaiser, 1998), FDM (Fang and Harrison, 2002), smooth-particle hydrodynamics (SPH) (Ma et al., 2011), cellular automaton (CA) (Feng et al., 2006), and lattice (Blair and Cook, 1998) models. Among these implementations, the realistic failure process analysis (RFPA) code of Tang and Kaiser (1998) can provide an effective description of microscopic damage mechanisms by assuming a Weibull distribution of the mechanical parameters, including Young's modulus and strength properties (Zhu et al., 2005). Application of RFPA to simulate the evolution of the EDZ around a circular opening was illustrated by Zhu and Bruhns (2008) and Wang et al. (2009), in the presence of material anisotropy and under hydro-mechanically coupled conditions, respectively.

Rock mass discontinuities can be explicitly incorporated into continuum models by means of discrete joint (or interface) elements. This technique, originally proposed byGoodman et al. (1968) and known as the combined continuum-interface method (Riahi et al., 2010), is however suited only for a relatively low number of discontinuities. Alternatively, if the number of discontinuities is large and the discontinuities are not preferably oriented, homogenization techniques can be employed. That is, the rock mass is modelled as continuum with reduced deformation and strength properties accounting for the degrading effect of local geological conditions (Hoek et al., 2002 and Hammah et al., 2008). Numerical homogenization of a continuum constitutive model can also be obtained from the results of discrete element simulations explicitly accounting for the presence of synthetic fracture networks (e.g. Beck et al., 2009).

Discontinuum approaches

In discrete (or discontinuous) modelling techniques, commonly known as the discrete element method (DEM), the material is treated as an assembly of independent, rigid or deformable blocks or particles. Unique features of the DEM are the abilities to: (i) capture finite displacements and rotations of discrete bodies, including complete detachment, and (ii) automatically recognise new contacts as the simulation progresses (Cundall and Hart, 1992). Unlike continuum methods, which are based on constitutive laws, DEM relies on interaction laws. Based on the different solution strategies, DEMs can be divided into two main groups (Jing and Stephansson, 2007). The first group, usually referred to as the distinct element method, uses an explicit time-domain integration scheme with finite difference discretization to solve the equations of motion for rigid or deformable discrete bodies with deformable contacts (Cundall and Strack, 1979). The most widely used codes of this type are the universal distinct element code (UDEC) (Itasca, 2013) for blocky systems and the particle flow code (PFC) (Itasca, 2012) for granular systems. The second category uses an implicit (and thus unconditionally stable) time integration scheme and it is represented mainly by the discontinuous deformation analysis (DDA) method (Shi and Goodman, 1988). Further classification of DEMs is based on criteria such as the type of contact between bodies, the representation of deformability of solid bodies, and the methodology for detection and revision of contacts (Jing and Stephansson, 2007).

While original applications of DEMs were mainly in the field of granular materials and jointed structures, further developments made DEMs also capable of explicitly simulating failure through intact rock material. Particularly, the concept of particle (or block) bonding, together with the introduction of cohesive contact models in DEMs, allowed the formation of new fractures to be captured. In this context, the FDEM (Munjiza, 2004 and Mahabadi, 2012) adopted is a special type of discontinuum approach, whereby the simulation effectively starts with a continuous representation of the solid domain and, as the simulation progresses with time, new discontinuities are allowed to form upon satisfying some fracture criterion, thus leading to the formation of new discrete bodies. For a detailed review of discrete methods, and their application to underground structures, the reader is referred to Lisjak and Grasselli (2014).

FUNDAMENTAL PRINCIPLES OF FDEM

The modelling platform adopted for the numerical simulations was the open source FDEM software known as Y-Geo (Mahabadi et al., 2012). In Y-Geo, the modelling domain is discretised with a mesh consisting of three-node triangular elements with four-node interface (or crack) elements embedded between the edges of all adjacent triangle pairs (Fig. 1a). The progressive failure of rocks is simulated using a cohesive-zone approach, a technique originally introduced in the context of the elasto-plastic fracturing of ductile metals (Dugdale, 1960) and then extended to quasi-brittle materials, such as concrete and rocks (Hillerborg et al., 1976). During elastic loading, stresses and strains are assumed to be distributed over the bulk material (i.e. the continuum portion of the model), which is therefore treated as linear-elastic using the triangular elements. Impenetrability between these elements is enforced by a penalty-based contact interaction algorithm (Munjiza and Andrews, 2000). Upon exceeding the peak strength of the material (in tension, shear, or a mixed-mode), the strains are assumed to localise within a narrow zone, known as the Fracture Process Zone (FPZ). The mechanical response of the FPZ is captured by a non-linear interdependence between stress and crack displacement implemented at the crack element level.

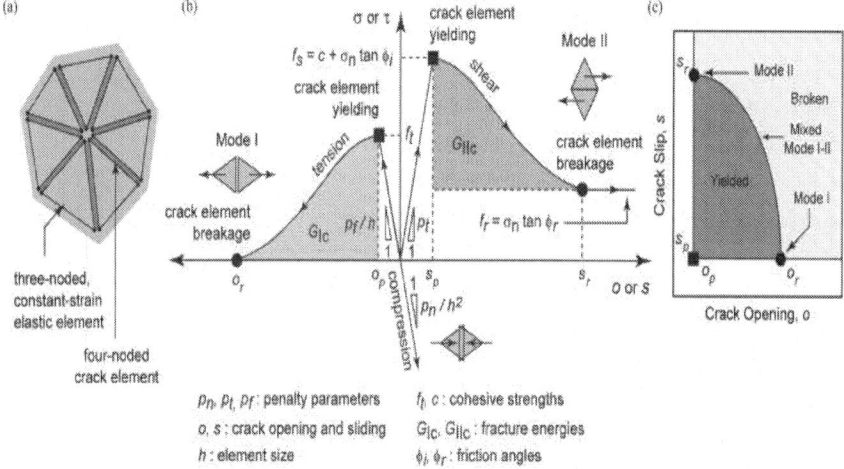

Figure 1. Simulation of rock deformation and fracturing with FDEM. (a) Representation of a continuum using cohesive crack elements interspersed throughout a mesh of triangular elastic elements. Triangles are shrunk for illustration purposes. (b) Constitutive behaviour of the crack elements defined in terms of normal and tangential bonding stresses, σ and τ, versus crack relative displacements, o and s (i.e. opening and slip). (c) Elliptical coupling relationship between o and s, for mixed-mode fracturing.

The constitutive response of a crack element is defined in terms of a variation of the bonding stresses, σ and τ, between the edges of the triangular element pair as a function of the crack relative displacements, o and s, in the normal and tangential directions, respectively (Fig. 1b). In tension (i.e. Mode I), the response of each crack element depends on the cohesive tensile strength, f_t, and the Mode I fracture energy, G_{Ic}. In shear (i.e. Mode II), the behaviour is governed by the peak shear strength, f_s, and the Mode II fracture energy, G_{IIc}. The peak shear strength is defined as

$$f_s = c + \sigma_n \tan \varphi_i \tag{1}$$

where c is the cohesion, φ_i is the internal friction angle, and σ_n is the normal stress acting across the crack element. The Mode I and Mode II fracture energies, G_{Ic} and G_{IIc}, represent the amount of energy, per unit crack length along the crack edge, consumed during the creation of a tensile and shear fracture, respectively. Upon breaking a crack element, a purely frictional resistance, f_r, is assumed to act along the newly-created discontinuity:

$$f_r = \sigma_n \tan \varphi_f \tag{2}$$

where φ_f is the fracture friction angle. For mixed Mode I–II fracturing, an elliptical coupling relationship is adopted between crack opening, o, and slip, s (Fig. 1c). Although no deformation should, in theory, occur in the crack elements before the cohesive strength is exceeded, a finite cohesive stiffness is required by the formulation of FDEM. Such an artificial stiffness is represented by the normal, tangential and fracture penalty values, p_n, p_t and p_f, for compressive, shear and tensile loading conditions, respectively. For practical purposes, the cohesive contribution to the overall model compliance can be largely limited by adopting very high (i.e. dummy) penalty values (Munjiza, 2004 and Mahabadi, 2012). Since fractures can nucleate only along the boundaries of the triangular elements, arbitrary fracture trajectories can be reproduced within the constraints imposed by the mesh topology. As the simulation progresses, through explicit time stepping, finite displacements and rotations of newly-created discrete bodies are allowed and new contacts are automatically recognised (Munjiza and Andrews, 1998).

The FDEM formulation described above was originally introduced to model isotropic materials. However, additional capabilities have been recently introduced into the FDEM solver to capture the mechanical response of anisotropic media (Lisjak et al., 2014a and Lisjak et al., 2014b). In particular, the modulus anisotropy is captured by a transversely isotropic, linear elastic constitutive law implemented at the triangular element level. In this case, the elastic deformation is fully characterised by five independent elastic parameters: two Young's moduli, E_P and E_S, and Poisson's ratios, v_P and v_S, for the directions parallel and perpendicular to the plane of isotropy, and the shear modulus, G_S. The anisotropy of strength is instead introduced at the crack element level by specifying the cohesive strength of each crack element as a function of the relative orientation, γ, between the crack element itself and the bedding orientation (Fig. 2a). The cohesive strength parameters and the fracture energies are assumed to vary linearly between a minimum value for $\gamma = 0°$ (i.e. $f_{t,min}$, c_{min}, $G_{Ic,min}$, $G_{IIc,min}$) to a maximum value for $\gamma = 90°$ (i.e. $f_{t,max}$, c_{max}, $G_{Ic,max}$, $G_{IIc,max}$). Furthermore, the mesh topology combines a random triangulation for the intra-layer material (i.e. matrix) together with crack elements preferably aligned along the plane of isotropy (Fig. 2b).

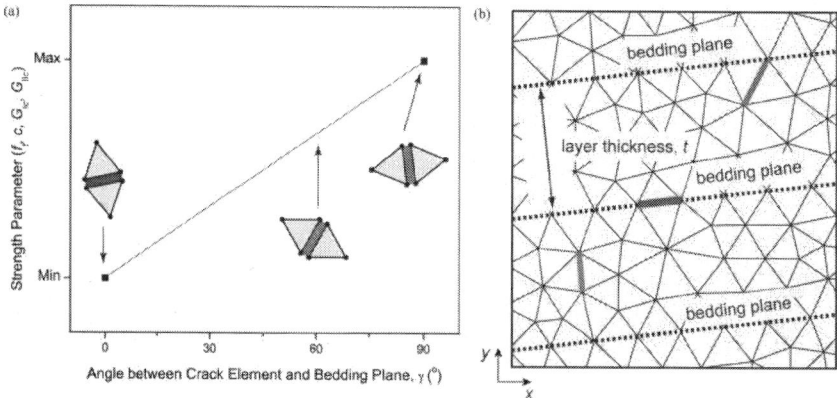

Figure 2. FDEM modelling of strength anisotropy. (a) Linear variation of cohesive strength parameters with the angle, γ, between crack element and layering orientation. (b) Example of mesh combining a Delaunay triangulation for the intra-layer material with edges preferentially aligned along the isotropy direction (afterLisjak et al. (2014a)).

CASE STUDIES

Model description

Geometries

Three different geomechanical scenarios were considered in the FDEM simulations: (i) a 6-m-diameter vertical shaft sunk in a horizontally bedded rock formation (shaft model,Fig. 3a), (ii) a 3-m-diameter tunnel excavated in a bedded formation (tunnel model,Fig. 3b), and (iii) two adjacent horseshoe-shaped caverns excavated in an isotropic rock (cavern model, Fig. 3c). The openings were placed at the centre of a square domain with dimensions equal to 100 m × 100 m, 50 m × 50 m, and 500 m × 500 m, for the shaft, tunnel, and cavern model, respectively. To maximise the model resolution in the EDZ, while keeping the run times within practical limits, a mesh refinement zone was adopted around the excavation boundaries, with an average element size of 0.06 m, 0.03 m, and 0.75 m, for the shaft, tunnel, and cavern model, respectively. The sensitivity of the model to variations in element size and topology was not investigated. In the tunnel model, the cross-section was assumed to be perpendicular to the strike of bedding planes inclined at $\psi = 33°$ from the horizontal. In the cavern model, five and four sub-domains were adopted for the powerhouse and transformer caverns, respectively, to analyse the effect of excavation staging on the fracturing process.

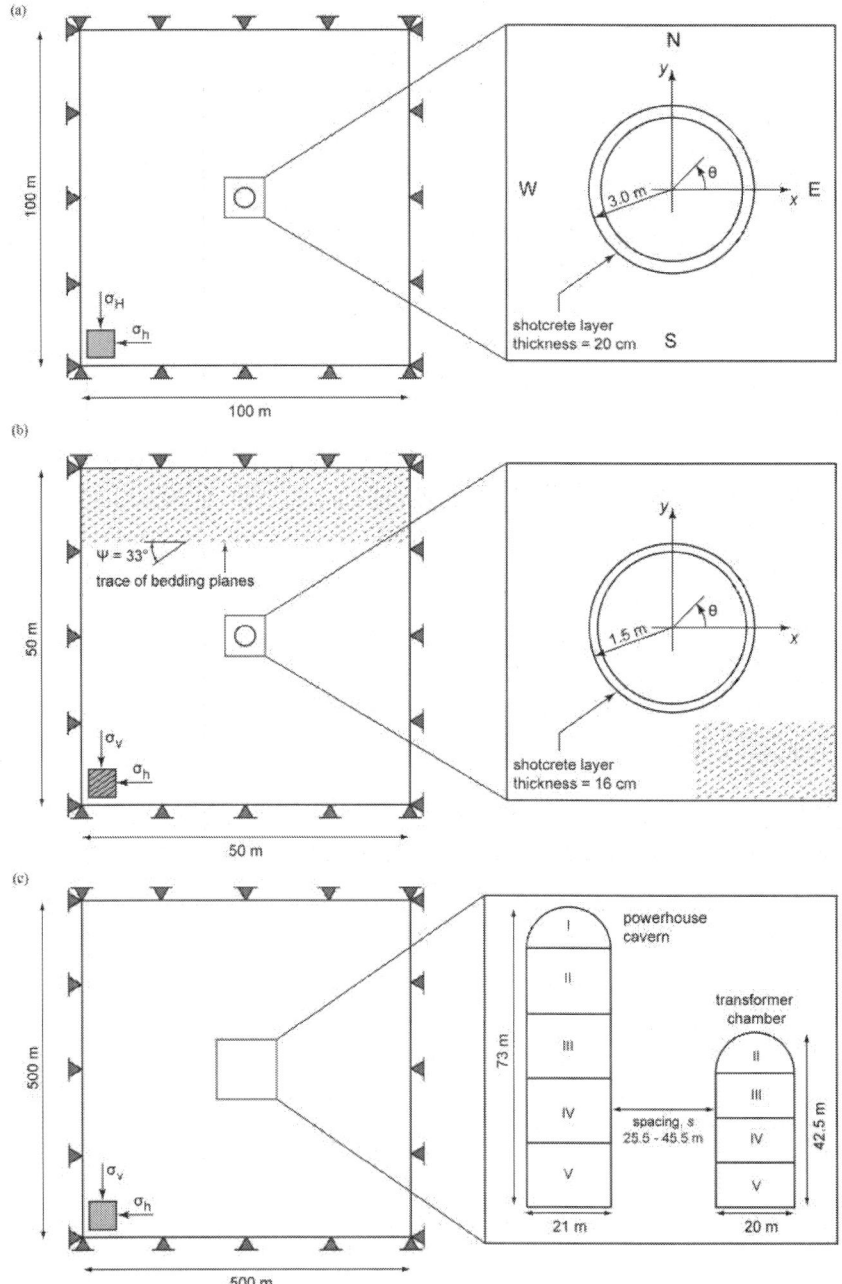

Figure 3. Geometry and boundary conditions of the FDEM models: (a) shaft, (b) tunnel, and (c) cavern.

In situ stresses and boundary conditions

To correctly simulate the prior-to-excavation stress state, each model required two separate runs. In the first run, the vertical and horizontal in situ stress conditions, as reported in Table 1, were applied to the model without the insertion of crack elements. Gravity-induced stress gradients were neglected. As suggested by Hudson and Harrison (1997), in the cavern model, three different in situ stress fields were simulated (stress ratio $K_0 = 0.5$, $K_0 = 1.0$ and $K_0 = 1.33$) in order to investigate the cases of pillar over-stressing (for $K_0 < 1$) as well as stress shadowing (for $K_0 > 1$). The first run was continued until the total kinetic energy of the system decayed to a negligible value (i.e. resulting stress waves were attenuated). The revised nodal coordinates corresponding to the system at rest (i.e. static equilibrium) were then obtained. Subsequently, these revised nodal coordinates were used as the current nodal coordinates (i.e. deformed mesh) of the second run in which the actual material strengths were assigned. By changing the far-field boundaries to be fixed in the horizontal and vertical directions, the first order in situ conditions were maintained while allowing the excavation to be initiated. Model relaxation induced by the artificial compliance of the crack elements was minimised by the choice of sufficiently large contact and fracture penalty values. It is noteworthy that only in-plane stresses were effectively used in the analysis, as the crack element formulation cannot account for the influence of an out-of-plane stress.

Table 1. Summary of in situ stress conditions applied to the excavation models. Vertical and horizontal stresses are oriented along the principal directions. The stress ratio, K_0, is reported in brackets.

Model	Vertical stress, σv (MPa)	Maximum horizontal stress, σH (MPa)	Minimum horizontal stress, σh (MPa)
Shaft		19.6	15.7 (0.8)
Tunnel	6.5		4.5 (0.7)
Cavern	8		4 (0.5)
	6		6 (1.0)
	6		8 (1.3)

Excavation and support modelling

With the correct in situ stress conditions achieved, the openings were created using a core replacement technique. With this approach, the three-dimensional supporting effect of the excavation face, which causes a gradual reduction of radial resistance around the excavation boundary, is captured by a fictitious, softening elastic material placed in the excavation core. In general, with this method, the deformation modulus of the excavated material is progressively reduced from the original rock mass value, corresponding to an undeformed section far ahead of the face, to a value that results in the wall displacements at the time of support installation. In this work, no attempt was made to match any real deformation measurements and, therefore, the modulus reduction ratio at the time of support activation was arbitrarily chosen. Since the deformation modulus of the excavation core was reduced over time in a stepwise fashion, the total kinetic energy of the model was again monitored to ensure that steady-state conditions were reached at every excavation stage. The final stage of the excavation sequence also involved the actual material removal and, for the shaft and tunnel models only, the activation of the support layer. To simplify the analysis, the effect of rock support was not considered in the cavern model. The application of shotcrete on the tunnel walls was modelled using constant-strain linear-elastic triangular elements. The support installation consisted of specifying the liner thickness and the installation time from a given core softening ratio. Since the delayed installation of shotcrete was accomplished by varying the elastic properties of the liner (from those of the rock mass to those of the shotcrete), the deformation in the liner had to be zeroed to avoid an artificial build-up of stress in response to an instantaneous increase of material stiffness in a pre-stressed medium.

Input parameters

Since the shaft was mined perpendicular to the layering strike, an isotropic mechanical model was assumed with input parameters based on laboratory values for an indurated claystone from Northern Switzerland (unpublished report) (Table 2). For the tunnel model, the rock mass was modelled using an anisotropic strength and stiffness model with a layering thickness of 0.1 m (Fig. 2). The input elastic properties as well as the cohesive strength parameters were those obtained from the back-analysis of a test tunnel excavated in an anisotropic shale formation (Opalinus Clay) at the Mont Terri underground research laboratory (URL) (Switzerland) (see Lisjak (2013) and Lisjak et al. (2014c) for further details). The rock mass parameters for the cavern model were based on unpublished laboratory values of a gneissic rock.

Table 2. FDEM input parameters of the three excavation models.

Input parameters	Shaft	Tunnel	Cavern
Continuum triangular elements			
Formulation type	Isotropic	Anisotropic	Isotropic
Bulk density, ρ (kg/m³)	2430	2430	2600
Young's modulus, E (GPa)	11.4		27
Young's modulus parallel to bedding, E_P (GPa)		3.8	
Young's modulus perpendicular to bedding, E_S (GPa)		1.3	
Poisson's ratio, v	0.27		0.3
Poisson's ratio parallel to bedding, v_P		0.35	
Poisson's ratio perpendicular to bedding, v_S		0.25	
Shear modulus, G_S (GPa)		3.6	
Viscous damping coefficient, μ (kg/(m s))	6.34×10^6	1.83×10^5	1.25×10^7

Crack elements			
Formulation type	Isotropic	Anisotropic	Isotropic
Tensile strength, f_t (MPa)	1.5		8
Tensile strength parallel to bedding, $f_{t,max}$ (MPa)		1.8	
Tensile strength perpendicular to bedding, $f_{t,min}$ (MPa)		0.44	
Cohesion, c (MPa)	12.9		4.5
Cohesion parallel to bedding, c_{min} (MPa)		2.8	
Cohesion perpendicular to bedding, c_{max} (MPa)		24.8	
Mode I fracture energy, G_{Ic} (J/m^2)	10.5		50
Mode I fracture energy parallel to bedding, $G_{Ic,max}$ (J/m^2)		19.5	
Mode I fracture energy perpendicular to bedding, $G_{Ic,min}$ (J/m^2)		1	
Mode II fracture energy, G_{IIc} (J/m^2)	105		200
Mode II fracture energy parallel to bedding, $G_{IIc,min}$ (J/m^2)		27.5	
Mode II fracture energy perpendicular to bedding, $G_{IIc,max}$ (J/m^2)		96.5	

Friction angle of intact material, φ_i (°)	24	22	39
Friction angle of fractures, φ_f (°)	24	22	30
Normal contact penalty, p_n (GPa m)	114	38	270
Tangential contact penalty, p_t (GPa/m)	11.4	3.8	27
Fracture penalty, p_f (GPa)	57	19	135

Fracturing process around a circular shaft

The simulation results of the shaft model highlight the stress-driven nature of the rock mass failure process under homogeneous and isotropic conditions. Upon reducing the elastic modulus of the tunnel core, the N–S-oriented in situ maximum principal stress flows around the shaft boundary, resulting in the development of a compressive stress concentration in the sidewalls. The intensity of this stress concentration is such that shear (i.e. Mode II) fractures start to nucleate (Fig. 4a). Due to the stress-free surface created by the excavation process, a state of unconfined (or moderately confined) compression arises in proximity to the shaft walls. Consequently, the failure mode closely resembles that observed for rock specimens subjected to uniaxial compressive stress. In agreement with the Mohr-Coulomb failure criterion, conjugate shear cracks tend to develop at angle of $45° \pm \varphi_i/2$ to the vertical compressive stress. As the shaft face advances (i.e. the core modulus is further reduced), the shear fractures tend to propagate away from the excavation and, at the same time, tend to curve and realign themselves in the direction of the far-field maximum principal stress (Fig. 4b, c). As a result, a characteristic fracture pattern consisting of multiple families of cracks resembling logarithmic-spiral rupture surfaces is created. The mutual intersection of these slip lines tends to break up the rock mass by forming distinct blocks and fragments. In close vicinity to the excavation walls, the occurrence of rock crushing and fine fragmentation is due to the higher stress concentration. At a distance from the excavation boundary, shearing of these fractures causes a local stress redistribution which tends to protect the intact rock. The activation of the shotcrete layer stabilises the rock fracturing process, thus allowing static equilibrium conditions to be reached (Fig. 4d). The final EDZ assumes an elliptical shape with major and minor axes oriented parallel to the minimum and maximum in situ principal stress directions, respectively (Fig. 5). The fractured zone extends for roughly 10 m (i.e. 1.7 times of shaft diameter) and 5 m (i.e. 0.8 times of shaft diameter) in the E–W and N–S directions, respectively. Arching of the maximum compressive stress, σ_1, can be observed around the damaged area, while the confining stress, σ_3, decreases to zero inside the EDZ. Overall, the simulated fracturing process is in good qualitative agreement with the mechanisms described by Barton (1993) and with experimental observations of borehole stability in massive isotropic rocks (e.g. Addis et al., 1990) and of excavation-induced fracture networks in a claystone formation (Armand et al., 2014).

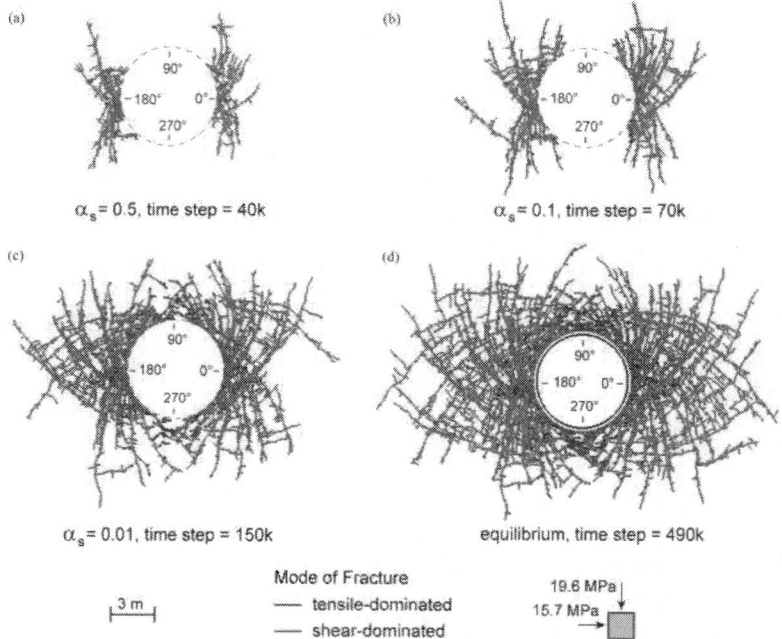

Figure 4. Shaft model: simulated evolution of fracture growth around the opening at increasing simulation times corresponding to different stages of the core modulus reduction sequence. The core modulus reduction ratio, a_s, is equal to the ratio of the core modulus to the rock mass modulus.

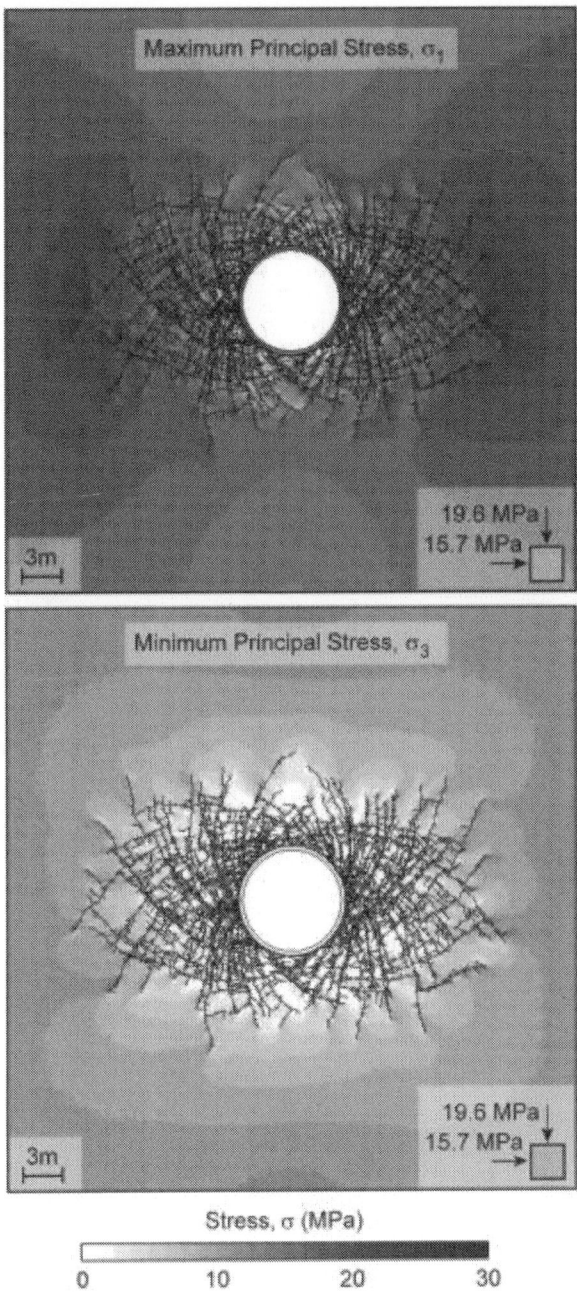

Figure 5. Shaft model: contours of maximum and minimum principal stresses associated with the excavation at equilibrium.

Influence of mechanical anisotropy on a circular tunnel
The numerical results of the tunnel model indicate that failure around an opening in a layered formation is triggered by the excavation-induced stress redistribution in combination with the lower strength of bedding planes favourably oriented for slip. As depicted in Fig. 6, fractures start to develop around the excavation boundary at approximately $0°\leq\theta\leq15°$, $120°\leq\theta\leq195°$, and $300°\leq\theta\leq360°$ in the form of shear-dominated (i.e. Mode II) fractures along the bedding direction. The polar orientation of these slip zones corresponds to critical values of relative orientation between compressive stress around the excavation boundary and bedding favourably oriented for slippage. As the simulation progresses, the slippage of bedding planes causes a local perturbation in the stress field which results in the nucleation of strain-driven, Mode I fractures in the direction perpendicular to the layering (Fig. 6b). Also, bedding-parallel sliding is simulated at about 70° and 250°.

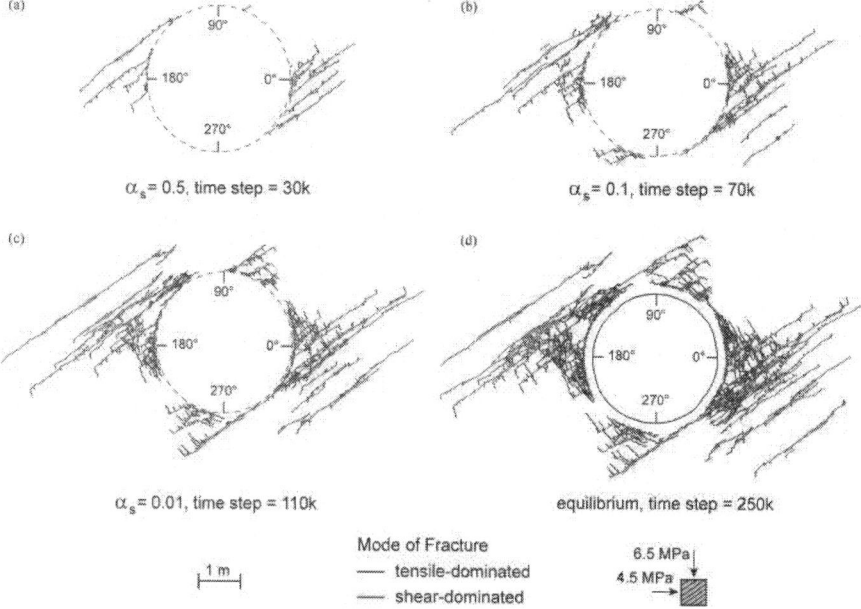

Figure 6. Tunnel model: simulated evolution of fracture growth around the tunnel at increasing simulation times corresponding to different stages of the core modulus reduction sequence.

Further rock mass deconfinement triggers further delamination of bedding planes (Fig. 6c) and the formation of wing-shaped fractured zones that tend to extend out in the direction parallel to the bedding to a distance of about 3 m from the sidewalls. After installing the support, the propagation of damage away from the opening is suppressed in favour of fragmentation in close proximity to the excavation boundary, until new equilibrium conditions are reached (Fig. 6d).

The total displacement field associated with the tunnel configuration at equilibrium (Fig. 7) indicates that at a distance from the excavation, the rock mass behaves elastically and, therefore, small strains, induced by the stress redistribution around the damaged zone, are simulated. Due to the highly anisotropic rock mass response, this distance varies from a minimum of 0.5 m to a maximum of 3 m in the direction parallel to bedding and in the sidewalls, respectively. Furthermore, elastic deformations of higher intensity are captured in the direction sub-perpendicular to the bedding orientation due to the high rock compressibility in the said direction. In the near-field excavation, an inner and an outer shell can be identified. The shape of the inner zone is roughly a 4.5 m × 4.5 m square with edges oriented in the direction parallel and perpendicular to the bedding and centre coincident with the tunnel axis. In this zone, the rock mass deformation is governed by a combination of Mode I and Mode II fracturing and bulking, thus resulting in large displacements (i.e. $\delta > 3$ cm). In the outer shell, while Mode II fractures can still nucleate, the relative sliding along the fracture surfaces is limited by higher values of confining stress, σ_3. Consequently, the growth of extensional fractures is effectively inhibited.

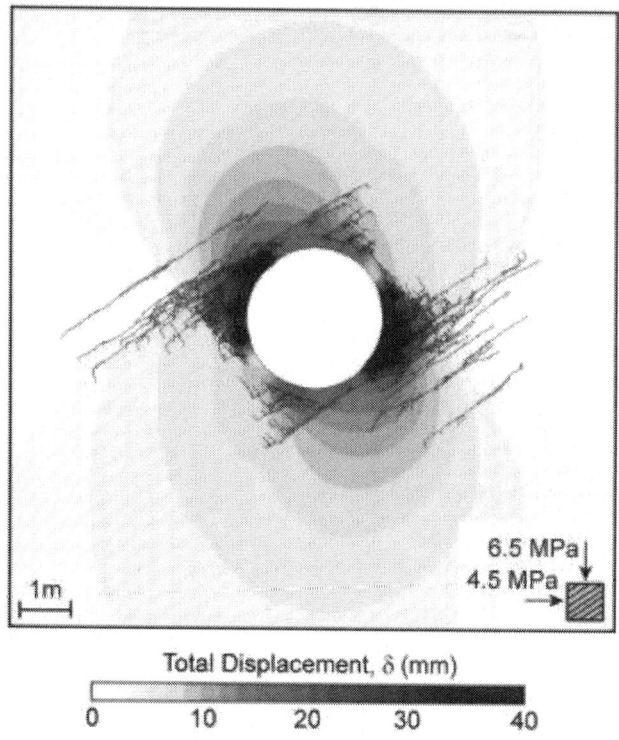

Figure 7. Tunnel model: colour contour of total displacement associated with the tunnel configuration at equilibrium.

The redistribution of compressive stress in response to the tunnel excavation (Fig. 8) is influenced by the in situ stress anisotropy as well as the characteristic fracture pattern (bedding-parallel discontinuities and a heavily fractured zone around the tunnel). The lateral extension of the EDZ due to bedding delamination is suppressed by the re-orientation of σ_1 in the direction perpendicular to bedding. In proximity to the tunnel boundary, bedding plane slippage promotes a drastic reduction of confining stress, σ_3, with low to moderate negative values responsible for the observed extensional fracturing.

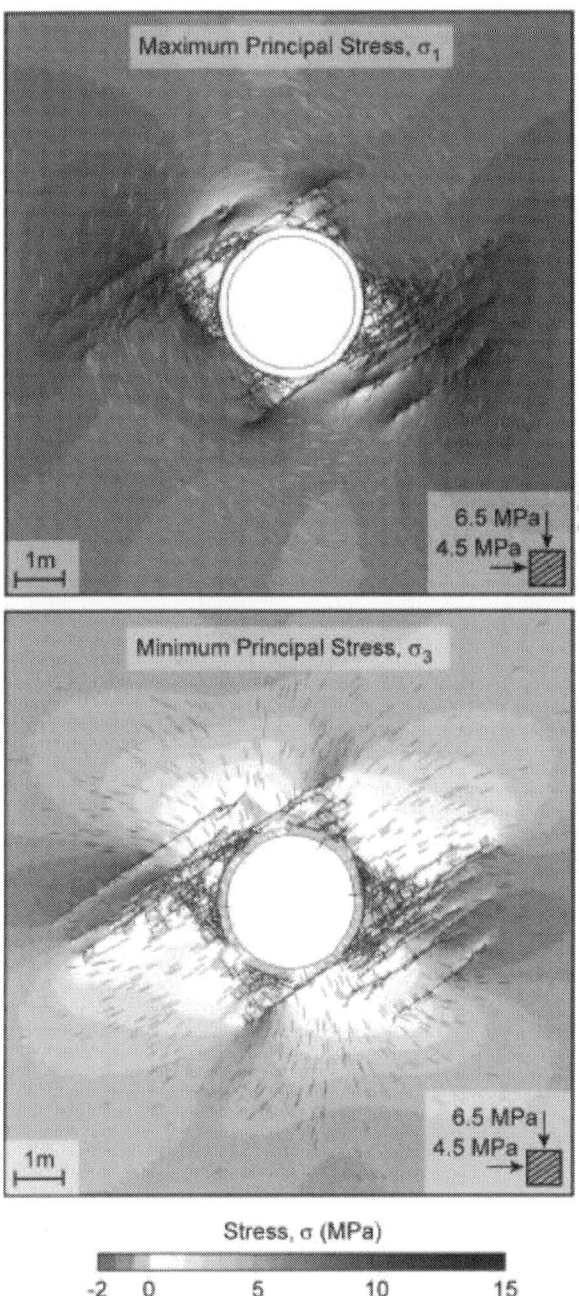

Figure 8. Tunnel model: colours contour of principal stresses associated with the tunnel configuration at equilibrium. Principal stress directions are indicated by short straight lines.

A quantitative comparison of the simulated damage pattern with specific in situ observations is beyond the scope of this work. Nevertheless, the failure mechanisms simulated here are in general agreement with a number of field and laboratory observations of excavations in laminated rock formations. In particular, the characteristic shear failure of bedding planes was observed in Opalinus Clay during hollow cylinder experiments (Labiouse and Vietor, 2014), and around boreholes, microtunnels and drifts at the Mont Terri URL (Marschall et al., 2006 and Blümling et al., 2007). Also, the importance of weakness planes in controlling the rock mass behaviour and the stability of underground openings is confirmed by observations from the construction of a hydroelectric tunnel in laminated sedimentary formations (Perras and Diederichs, 2009). Characteristic square-shaped fractured zones have also been reported in the hydrocarbon exploration industry when drilling horizontal boreholes in laminated shales (e.g. Økland and Cook, 1998 and Willson et al., 1999).

Interaction mechanisms between two adjacent caverns
Effect of in situ stress

The numerical results indicate a critical influence of the in situ stresses on the fracture development around the two caverns (Fig. 9). In general, for the adopted rock mass properties and cavern configuration, an isotropic stress field induces the lowest deviatoric stresses in the surrounding rock mass and, therefore, minimises damage development in the pillar. On the other hand, a vertically oriented maximum in situ principal stress leads to pillar over-stressing and failure. Also, higher stress concentrations tend to occur, in agreement with the analysis of Brady and Brown (2006), around high-curvature boundaries, which therefore become preferential loci of fracture initiation.

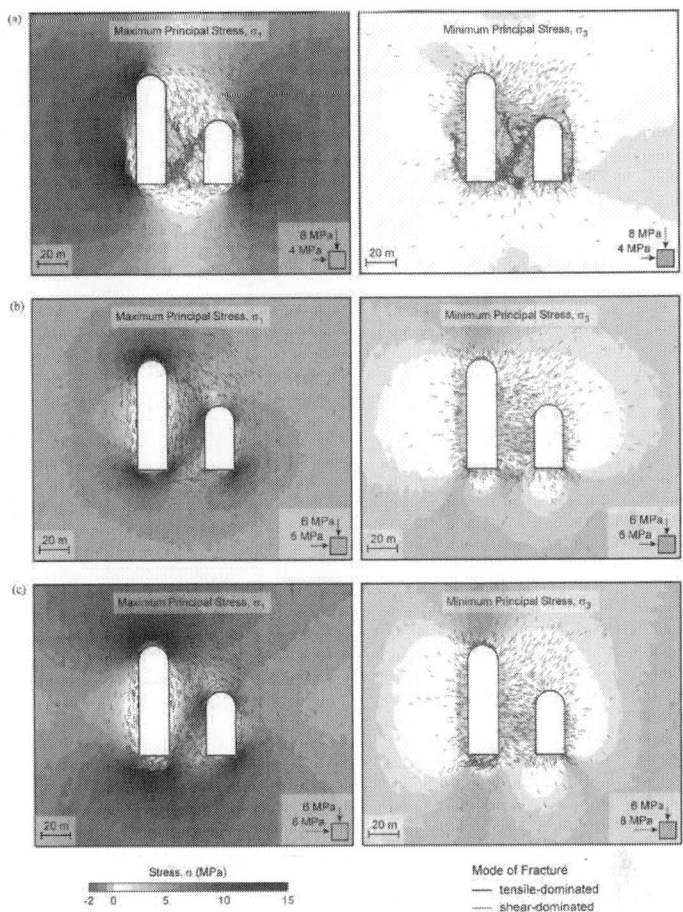

Figure 9. Effect of in situ stress anisotropy in the cavern model. Final stress distribution and fracture pattern for the cases of (a) $K_0 = 0.5$, (b) $K_0 = 1.0$, and (c) $K_0 = 1.33$. Colour contours on the left and right hand side represent the maximum and minimum principal stresses, σ_1 and σ_3, respectively. Local principal stress directions are indicated by short straight lines.

For the case of $K_0 = 0.5$ (Fig. 9a), the stress channelling within the rock pillar causes the development of a through-going macroscopic fracture plane, resembling that often observed in rock specimens subjected to uniaxial compression. The EDZ starts to form with a fracture growing from the lower right corner of the powerhouse cavern to the upper left corner of the transformer chamber. The stress redistribution causes further fracturing to originate from the centre of the pillar and propagate towards the upper right sidewall of the powerhouse and the lower left sidewall of the transformer cavern. Although a low to slightly negative confining stress, σ_3, develops in the rock pillar, the fracturing process is dominated by Mode II failure, due to the relatively high tensile strength of the rock compared to the cohesion value (Table 1).

Under isotropic stress conditions ($K_0 = 1.0$, Fig. 9b), damage develops around the lower corners of the two excavations and above the arched roof of the powerhouse cavern. Unlike the previous case ($K_0 = 0.5$), the rock pillar remains substantially intact. For the case of $K_0 = 1.33$ (Fig. 9c), the simulated failure process leads to an EDZ network that is overall similar to the isotropic case. One notable difference is represented by the larger fractured areas simulated in the back and roof of the powerhouse cavern due to the less favourable orientation of the in situ stress field.

Effect of pillar width

As described by Hoek (2006), the distance between the two caverns should be as small as possible to minimise the length of the busbars that connect the generators in the powerhouse cavern to the transformers in the adjacent cavern. On the other hand, this distance has to be large enough to preserve the structural integrity of the pillar. Therefore, the optimisation of the pillar width represents a crucial aspect of the design of this type of underground structures. In this study, the pillar damage was simulated for cavern spacing values, s, of 25.5 m, 35.5 m and 45.5 m, corresponding to ratios of the cavern width to the pillar width equal to 1.2, 1.7 and 2.2, respectively. Only the case of $K_0 = 0.5$ was analysed. As expected, the extent of the damaged area decreases as the spacing increases (Fig. 10). For $s = 25.5$ m (Fig. 10a), the EDZ spans the entire width of the pillar and is characterised by heavy fragmentation developing between the two caverns (see also Section 4.4.1). For $s = 35.5$ m, the damaged zones of the two caverns are still inter-connected (Fig. 10b), however fracturing is sensibly less intense than the previous case and a narrow load-bearing zone is preserved at the centre of the pillar. Also, unlike the first case, the stress concentration along the external sidewalls is too low to induce significant fracturing. Lastly, for $s = 45.5$ m (Fig. 10c), although a noticeable disturbance of the pre-excavation stress field is still presented, the stress redistribution does not cause a connected failure pattern to develop. Instead, an asymmetric EDZ fracture network is created: fracturing is concentrated along the inner sidewall of the transformer cavern, while, interestingly, the rock mass around the powerhouse cavern remains nearly intact.

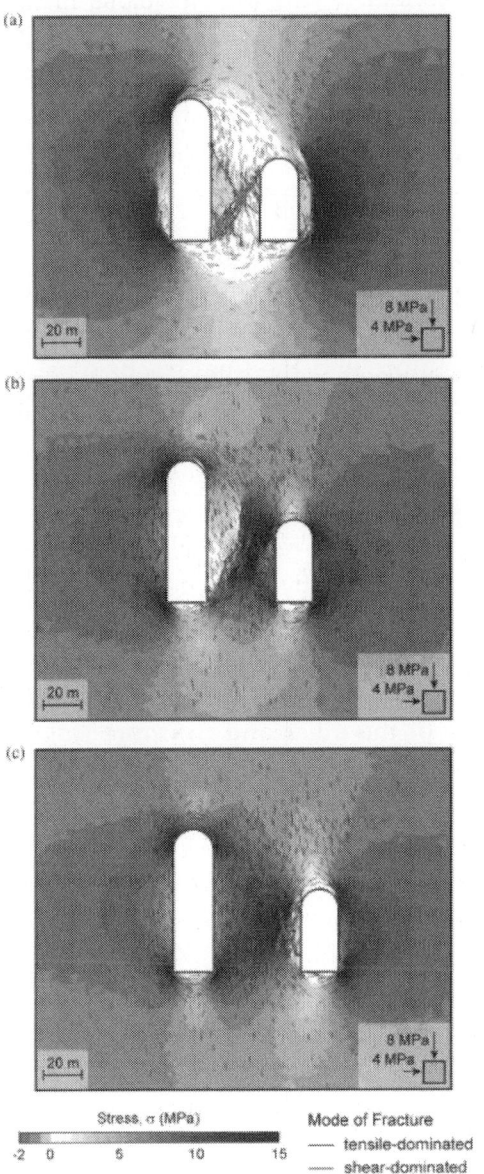

Figure 10. Effect of pillar width in the cavern model. Final stress distribution and fracture pattern are presented for a cavern spacing of (a) 25.5 m, (b) 35.5 m, and (c) 45.5 m. Colour contours represent the maximum principal stresses, σ_1. Local major principal stress directions are indicated by short straight lines.

Effect of excavation staging

The excavation of the entire cavern cross-section at once arguably represents an extreme case, which leads, as described above for $K_0 = 0.5$, to a large over-stressed and fractured area within the rock pillar. To investigate the adoption of a more realistic excavation procedure, further simulations were carried out, for the case of $K_0 = 0.5$ and $s = 35.5$ m, whereby both caverns were excavated using a multi-stage sequence. The excavation process started with the top heading of the powerhouse cavern (stage I), followed by a sequential excavation of the remaining four sub-domains of both caverns (stages II–V). The simulation results indicate that the stress history plays an important role in controlling the evolution of failure and the final fracture pattern (Fig. 11). For the one-stage excavation (Fig. 10b), fractures initiate from the bottom right corner of the powerhouse cavern and propagate towards the back of the transformer cavern. In contrast, in the case of a staged excavation, the failure process starts from the upper right sidewall of the powerhouse cavern (Fig. 11a). The material removal is accompanied by the downward growth of shear fractures sub-parallel to the caverns' inner sidewalls (Fig. 11b, c). Compared to the single-stage simulation, the final fracture pattern of the multi-stage model (Fig. 11d) shows (i) a larger residual intact area in the centre of the pillar and (ii) more fragmentation in close proximity to the sidewalls, due to a repeated occurrence of high compressive stresses around the corners of the excavation benches.

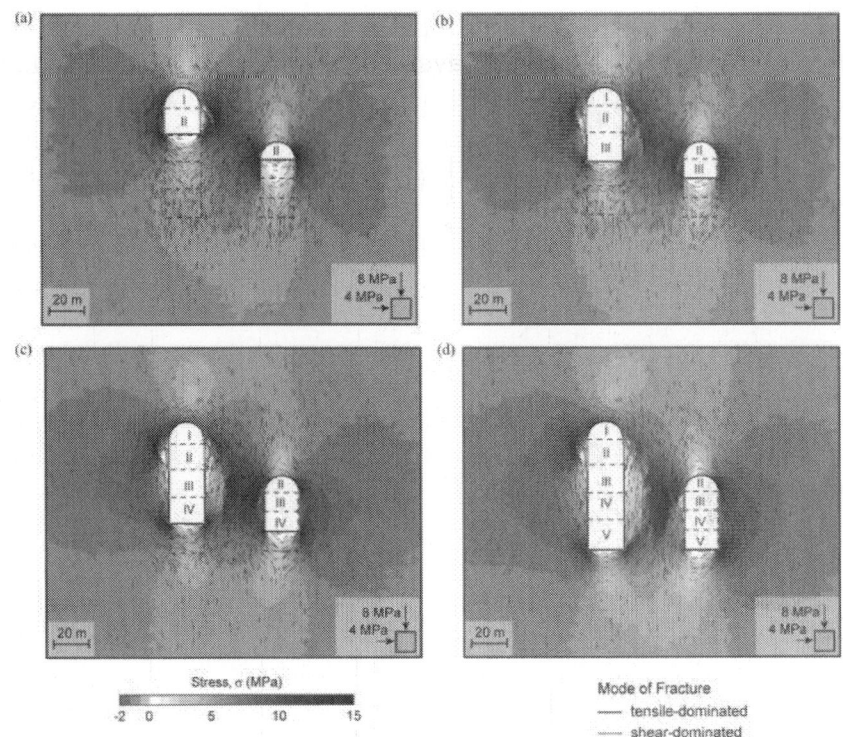

Figure 11. Effect of excavation staging in the cavern model. Failure sequence and maximum principal stress contour within the rock pillar for the case of $K_0 = 0.5$ and spacing of 35.5 m after (a) stage II, (b) stage III, (c) stage IV, and (d) stage V. Local major principal stress directions are indicated by short straight lines.

CONCLUDING REMARKS

Hybrid continuum-discontinuum simulations, based on the FDEM, were used to investigate excavation-induced fracturing processes around different types of underground structures. Three main geomechanical scenarios were considered.

Firstly, for a circular shaft excavated in a homogeneous and isotropic medium, shear failure started in the regions of the highest excavation-induced compressive stress concentration. Subsequently, fractures tended to follow characteristic trajectories, resembling the logarithmic-spiral slip zones captured by conventional Mohr-Coulomb elasto-plastic models, and in agreement with experimental observations from borehole breakout experiments.

Secondly, in the case of a tunnel excavated in a laminated shale, the bedding induced mechanical anisotropy was shown to sensibly influence the locus of fracture initiation as well as the direction of fracture growth. Due to the lower shear strength along the bedding, the rock failure was strongly dependent upon the relative orientation between bedding planes and in situ principal stress directions. Furthermore, the initial shearing along bedding planes induced a tensile stress state in the perpendicular direction with consequent formation of secondary strain-driven extensional fractures.

The third scenario focused on the behaviour of two adjacent horseshoe-shaped caverns. For the case of a vertically oriented in situ maximum principal stress, pillar over-stressing with formation of a through-going shear fracture plane was simulated, whereas isotropic in situ stress conditions resulted in the least amount of rock damage. A sensitivity analysis to the pillar width revealed that, to avoid the formation of an interconnected EDZ between the two caverns and, therefore, preserve the pillar load-bearing capacity; the cavern spacing should be greater than about two times the cavern width. Finally, the adoption of a multi-stage excavation sequence was shown to affect the fracture growth as well as the final damage pattern.

In conclusion, the modelling results indicate that FDEM simulations can provide unique geomechanical insights in all those cases where an explicit consideration of fracture and fragmentation processes is of paramount importance.

ACKNOWLEDGEMENTS

This work has been supported by the Natural Science and Engineering Research Council (NSERC) of Canada in the form of discovery grant No. 341275 and the Swiss National Cooperative for the Disposal of Radioactive Waste (NAGRA).

REFERENCES

1. Addis MA, Barton NR, Bandis SC, Henry JP. Laboratory studies on the stability of vertical and deviated boreholes. In: Proceedings of the SPE Annual Technical Conference and Exhibition, New Orleans, Louisiana; September 23e26, 1990.

2. Armand G, Leveau F, Nussbaum C, de La Vaissiere R, Noiret A, Jaeggi D, Landrein P, Righini C. Geometry and properties of the excavation-induced fractures at the Meuse/Haute-Marne URL drifts. Rock Mechanics and Rock Engineering 2014;47(1):21e41.

3. Barton NR. From empiricism, through theory, to problem solving in rock engineering. In: Harmonising rock engineering and the Environment. London, UK: CRC Press; 2011. p. 3e16.

4. Barton NR. Physical and discrete element models of excavation and failure in jointed rock. In: Keynote lecture presented at ISRM International Symposium on assessment and prevention of failure phenomena in rock engineering, Istanbul, Turkey; 1993.

5. Bazant ZP, Pijaudier-Cabot G. Nonlocal continuum damage, localization instability and convergence. Journal of Applied Mechanics 1988;55(2):287e93.

6. Beck DA, Pfitzner MJ, Arndt SM, Fillery B. Estimating rock mass properties and seismic response using higher order, discontinuous, finite element models. In: Proceedings of the 3rd Canada-US Rock Mechanics Symposium. Toronto, Canada: University of Toronto Press; 2009. Paper 4189.

7. Belytschko T, Möes N, Usui S, Parimik C. Arbitrary discontinuities in finite elements. International Journal for Numerical Methods in Engineering 2001;50(4):993e 1013.

8. Blair S, Cook NGW. Analysis of compressive fracture in rock using statistical techniques: part I. A non-linear rule-based model. International Journal of Rock Mechanics and Mining Sciences 1998;35(7):837e48.

9. Blümling P, Bernier F, Lebon P, Martin CD. The excavation damaged zone in clay formations time-dependent behaviour and influence on performance assessment. Physics and Chemistry of the Earth, Parts A/B/C 2007;32(8e14):588e 99.

10. Brady BHG, Brown ET. Rock mechanics for underground mining. Dordrecht, Netherlands: Springer; 2006.

11. Cai M, Kaiser PK. In situ rock spalling strength near excavation boundaries. Rock Mechanics and Rock Engineering 2014;47(2):659e75.

12. Cundall PA, Hart RD. Numerical modelling of discontinua. Engineering Computations 1992;9(2):101e13.

13. Cundall PA, Strack ODL. A discrete numerical model for granular assemblies. Geotechnique 1979;29(1):47e65.

14. de Borst R, Sluis LJ, Mühlhaus HB, Pamin J. Fundamental issues in finite element analyses of localization of deformation. Engineering Computations 1993;10(2): 99e122.

15. Deb D, Das KC. Extended finite element method for the analysis of discontinuities in rock masses. Geotechnical and Geological Engineering 2010;28(5):643e59.

16. Dugdale DS. Yielding of steel sheets containing slits. Journal of the Mechanics and Physics of Solids 1960;8(2):100e4

17. Eberhardt E. Numerical modelling of three-dimension stress rotation ahead of an advancing tunnel face. International Journal of Rock Mechanics and Mining Sciences 2001;38(4):499e518.

18. Fang Z, Harrison JP. Development of a local degradation approach to the modelling of brittle fracture in heterogeneous rocks. International Journal of Rock Mechanics and Mining Sciences 2002;39(4):443e57.

19. Feng XT, Pan PZ, Zhou H. Simulation of the rock microfracturing process under uniaxial compression using an elasto-plastic cellular automaton. International Journal of Rock Mechanics and Mining Sciences 2006;43(7):1091e108.

20. Goodman RE, Taylor RL, Brekke TA. A model for the mechanics of jointed rock. Journal of the Soil Mechanics and Foundations Division 1968;94(3):637e60.

21. Hammah RE, Yacoub T, Corkum B, Curran JH. The practical modelling of discontinuous rock masses with finite element analysis. In: Proceedings of the 42nd US rock mechanics Symposium and 2nd US-Canada rock mechanics Symposium, San Francisco, USA. Alexandria, Virginia, USA: American Rock Mechanics Association; 2008. ARMA 08e180.

22. Hillerborg A, Modeer M, Petersson PE. Analysis of crack formation and crack growth in concrete by means of fracture mechanics and finite elements. Cement and Concrete Research 1976;6(6):773e81.

23. Hoek E, Carranza-Torres CT, Corkum B. Hoek-Brown failure criterion e 2002 edition. In: Proceedings of the 5th North American rock mechanics Symposium. Toronto, Canada: University of Toronto Press; 2002. p. 267e73.

24. Hoek E, Kaiser PK, Bawden WF. Support of underground excavations in hard rock. Leiden, Netherlands: Taylor & Francis/Balkema; 1995.

25. Hoek E. Practical rock engineering. 2006. http://www.rocscience.com/hoek/pdf/ Practical_Rock_Engineering.pdf. Hudson JA, Harrison JP. Engineering rock mechanics, vol. I. Pergamon; 1997.

26. Itasca. PFC2D (particle flow code in 2 dimensions). Minneapolis, MN, USA: Itasca Consulting Group Inc.; 2012. Itasca. UDEC (universal distinct element code). Minneapolis, MN, USA: Itasca Consulting Group Inc.; 2013.

27. Jaeger JC, Cook NGW, Zimmerman RW. Fundamentals of rock mechanics. 4th ed. Malden, MA, USA: Blackwell Publishing; 2007. Jing L, Hudson JA. Numerical methods in rock mechanics. International Journal of Rock Mechanics and Mining Sciences 2002;39(4):409e27.

28. Jing L, Stephansson O. Fundamentals of discrete element methods for rock engineering: theory and applications, volume 85 (developments in geotechnical engineering). Amsterdam, Netherlands: Elsevier; 2007.

29. Karekal S, Das R, Mosse L, Cleary PW. Application of a mesh free continuum method for simulation of rock caving processes. International Journal of Rock Mechanics and Mining Sciences 2011;48(5):703e11.

30. Labiouse V, Vietor T. Laboratory and in situ simulation tests of the excavation damaged zone around galleries in Opalinus Clay. Rock Mechanics and Rock Engineering 2014;47(1):57e70.

31. Lisjak A, Grasselli G. A review of discrete modelling techniques for fracturing processes in discontinuous rock masses. Journal of Rock Mechanics and Geotechnical Engineering 2014;6(4):301e14.

32. Lisjak A, Grasselli G, Vietor T. Continuum-discontinuum analysis of failure mechanisms around unsupported circular excavations in anisotropic clay shales. International Journal of Rock Mechanics and Mining Sciences 2014a;65:96e 115.

33. Lisjak A, Tatone BSA, Grasselli G, Vietor T. Numerical modelling of the anisotropic mechanical behaviour of Opalinus Clay at the laboratory scale using FEM/DEM. Rock Mechanics and Rock Engineering 2014b;47(1):187e206.

34. Lisjak A, Garitte B, Grasselli G, Müller H, Vietor T. The excavation of a circular tunnel in a bedded argillaceous rock (Opalinus Clay): short-term rock mass response and numerical analysis using FDEM. Tunnelling and Underground Space Technology 2014c. http://dx.doi.org/10.1016/j.tust.2014.09.014 (in press).

35. Lisjak A. Investigating the influence of mechanical anisotropy on the fracturing behaviour of brittle clay shales with application to deep geological repositories. PhD Thesis. Toronto, Canada: University of Toronto; 2013., http://hdl.handle. net/1807/43649.

36. Ma GW, Wang XJ, Ren F. Numerical simulation of compressive failure of heterogeneous rock-like materials using SPH method. International Journal of Rock Mechanics and Mining Sciences 2011;48(3):353e63.

37. Mahabadi OK, Lisjak A, Grasselli G, Munjiza A. Y-Geo: a new combined finitediscrete element numerical code for geomechanical applications. International Journal of Geomechanics 2012;12(6):676e88.

38. Mahabadi OK. Investigating the influence of micro-scale heterogeneity and microstructure on the failure and mechanical behaviour of geomaterials. Ph.D. thesis. Toronto, Canada: University of Toronto; 2012., http://hdl.handle.net/ 1807/32789.

39. Marschall P, Distinguin M, Shao H, Bossart P, Enachescu C, Trick T. Creation and evolution of damage zones around a microtunnel in a claystone formation of the Swiss Jura Mountains. In: Proceedings of the International Symposium and Exhibition on formation damage Control. Lafayette, Louisiana, USA: Society of Petroleum Engineers; 2006. Martin CD. Seventeenth Canadian geotechnical colloquium: the effect of cohesion loss and stress path on brittle rock strength. Canadian Geotechnical Journal 1997;34(5):239e54.

40. Masin D. A hypoplastic constitutive model for clays. International Journal for Analytical and Numerical Methods in Geomechanics 2005;29(4):311e36.

41. Mizukoshi T, Mimaki Y. Deformation behaviour of a large underground cavern. Rock Mechanics and Rock Engineering 1985;18(4):227e51.

42. Möes N, Belytschko T. Extended finite element method for cohesive crack growth. Engineering Fracture Mechanics 2002;69(7):773e81.

43. Mühlhaus HB, Vardoulakis I. The thickness of shear bands in granular materials. Geotechnique 1987;37(3):845e57.

44. Munjiza A, Andrews KRF. NBS contact detection algorithm for bodies of similar size. International Journal for Numerical Methods in Engineering 1998;43(1):131e 49.

45. Munjiza A, Andrews KRF. Discretised penalty function method in combined finitediscrete element analysis. International Journal for Numerical Methods in Engineering 2000;49(12):1495e520.

46. Munjiza A. The combined finite-discrete element method. Chichester, West Sussex, England: John Wiley & Sons Ltd; 2004.

47. Økland D, Cook JM. Bedding-related borehole instability in high angle wells. In: Proceedings of the SPE/ISRM rock mechanics in Petroleum engineering. Trondheim, Norway: Society of Petroleum Engineers; 1998.

48. Peiro J, Sherwin S. Finite difference, finite element and finite volume methods for partial differential equations, chapter 8.2. Dordrecht, Netherlands: Springer; 2005.

49. Perras MA, Diederichs MS. Tunnelling in horizontally laminated ground. In: Diederichs MS, Grasselli G, editors. Proceedings of 3rd Canada-US (CANUS) rock mechanics Symposium (RockEng09), Toronto, Canada; 2009.

50. Rabczuk T, Belytschko T. A three-dimensional large deformation meshfree method for arbitrary evolving cracks. Computer Methods in Applied Mechanics and Engineering 2007;196(29e30):2777e99.

51. Rabczuk T, Belytschko T. Cracking particles: a simplified meshfree method for arbitrary evolving cracks. International Journal for Numerical Methods in Engineering 2004;61(13):2316e43.

52. Riahi A, Hammah ER, Curran JH. Limits of applicability of the finite element explicit joint model in the analysis of jointed rock problems. In: Proceedings of the 44th American rock mechanics Association (ARMA) Symposium; 2010. Shi G, Goodman RE. Discontinuous deformation analysis - a new method for computing stress, strain and sliding of block systems. In: Proceedings of the 29th US Symposium on Rock Mechanics. Minneapolis, USA: A.A. Balkema; 1988. p. 381e93.

53. Steer P, Cattina R, Lave J, Godardd V. Surface Lagrangian remeshing: a new tool for studying long term evolution of continental lithosphere from 2D numerical modelling. Computers and Geosciences 2011;37(8):1067e74.

54. Strouboulis T, Babuska I, Copps K. The design and analysis of the generalized finite element method. Computer Methods in Applied Mechanics and Engineering 2000;181(1e3):43e69.

55. Tang CA, Kaiser PK. Numerical simulation of cumulative damage and seismic energy release during brittle rock failure e Part I: fundamentals. International Journal of Rock Mechanics and Mining Sciences 1998;35(2):113e21.

56. Wang SH, Lee CI, Ranjith PG, Tang CA. Modelling the effects of heterogeneity and anisotropy on the excavation damaged/disturbed zone (EDZ). Rock Mechanics and Rock Engineering 2009;42(2):229e58.

57. Willson SM, Last NC, Zoback MD, Moos D. Drilling in South America: a wellbore stability approach for complex geologic conditions. In: Proceedings of the Latin American and Caribbean Petroleum engineering Conference. Caracas, Venezuela: Society of Petroleum Engineers; 1999.

58. Zhang YL, Feng XT. Extended finite element simulation of crack propagation in fractured rock masses. Materials Research Innovations 2011;15:594e6.

59. Zhu WC, Bruhns OT. Simulating excavation damaged zone around a circular opening under hydromechanical conditions. International Journal of Rock Mechanics and Mining Sciences 2008;45(5):815e30.

60. Zhu WC, Liu J, Tang CA, Zhao XD, Brady BH. Simulation of progressive fracturing processes around underground excavations under biaxial compression. Tunnelling and Underground Space Technology 2005;20(3):231e47.

61. Zhuang X, Augarde CE, Mathisen KM. Fracture modelling using meshless methods and level sets in 3D: framework and modelling. International Journal for Numerical Methods in Engineering 2012;92(11):969e98.

62. Zhuang X, Zhu H, Augarde C. An improved meshless Shephard and least square method possessing the delta property and requiring no singular weight function. Computational Mechanics 2014;53(2):343e57.

CITATION

Andrea Lisjak, Daniel Figi, Giovanni Grasselli, Fracture development around deep underground excavations: Insights from FDEM modelling, Journal of Rock Mechanics and Geotechnical Engineering, Volume 6, Issue 6, December 2014, Pages 493-505, ISSN 1674-7755, http://dx.doi.org/10.1016/j.jrmge.2014.09.003.

CHAPTER 2

A Review of Discrete Modeling Techniques for Fracturing Processes in Discontinuous Rock Masses

A. Lisjak, G. Grasselli

Department of Civil Engineering, University of Toronto, Toronto M5S 1A4, Canada

ABSTRACT

The goal of this review paper is to provide a summary of selected discrete element and hybrid finite–discrete element modeling techniques that have emerged in the field of rock mechanics as simulation tools for fracturing processes in rocks and rock masses. The fundamental principles of each computer code are illustrated with particular emphasis on the approach specifically adopted to simulate fracture nucleation and propagation and to account for the presence of rock mass discontinuities. This description is accompanied by a brief review of application studies focusing on laboratory-scale models of rock failure processes and on the simulation of damage development around underground excavations.

INTRODUCTION

A large body of experimental research shows that the failure process in brittle rocks under compression is characterized by complicated micromechanical processes, including the nucleation, growth and coalescence of microcracks, which lead to strain localization in the form of macroscopic fracturing (Lockner et al., 1991 and Benson et al., 2008). The evolution of micro-cracking, typically associated with the emission of acoustic energy (AE), results in a distinctive non-linear stress–strain response, with macroscopic strain softening commonly observed under low-confinement conditions (Brace et al., 1966, Bieniawski, 1967, Eberhardt et al., 1997 and Martin, 1997). Furthermore, unlike other materials (e.g. metals), rocks exhibit a strongly pressure-dependent mechanical behavior (Jaeger and Cook, 1976). A variation of failure mode, from axial splitting to shear band formation, is indeed often observed for increasing confining pressures (Horii and Nemat-Nasser, 1986). This variation of failure behavior is reflected in a non-linear failure envelope (Kaiser and Kim, 2008) and a transition from brittle to ductile post-peak response (Paterson and Wong, 2004). At rock mass level, the failure process observed during laboratory-scale experiments is further complicated by the presence of discontinuities, such as joints, faults, shear zones, schistosity planes, and bedding planes (Goodman, 1989). Specifically, discontinuities affect the response of the intact rock by reducing its strength and inducing non-linearities and anisotropy in the stress–strain response (Hoek, 1983 and Hoek et al., 2002). Furthermore, discontinuities add kinematic constraints on the deformation and failure modes of structures in rocks (Hoek et al., 1995 and Hoek, 2006) and cause stress and displacement redistributions to sensibly deviate from linear elastic, homogenous conditions (Hammah et al., 2007).

Aside from the intrinsic uncertainties associated with the determination of reliable *in situ* input parameters, the application of numerical modeling to the analysis of rock engineering problems represents a challenging task owing to the aforementioned features of the rock behavior. In particular, the progressive degradation of material integrity during the deformation process, together with the influence of pre-existing discontinuities on the rock mass response, has represented a major drive for the development of new modeling techniques. In this context, the available numerical approaches are typically classified either as continuum- or discontinuum-based methods (Jing and Hudson, 2002).

The main assumption of continuum-based methods is that the computational domain is treated as a single continuous body. Standard continuum mechanics formulations are based on theories such as plasticity and damage mechanics, which adopt internal variables to capture the influence of history on the evolution of stress and changes at the micro-structural level, respectively (De Borst et al., 2012). Conventionally, the implementation of continuum techniques is based on numerical methods, such as non-linear finite element method (FEM), Lagrangian finite difference method (FDM), and boundary element method (BEM), with the incorporation of plasticity-based material models. However, standard strength-based strain-softening constitutive relationships cannot capture localization of failure as the lack of an internal length scale results in the underlying mathematical problem to become ill-posed (De Borst et al., 1993). Among the main consequences of adopting a standard continuum to simulate strain localization is the fact that, by doing so, localization occurs in a region of zero thickness and consequently an unphysical mesh sensitivity arises. To overcome these shortcomings, the description of the continuum must account either for the viscosity of the material, by incorporating a deformation-rate dependency, or for the change in the material micro-structure, by enhancing the mathematical formulation with additional terms (De Borst et al., 1993). The latter technique, known as regularization, includes non-local (e.g. Bažant and Pijaudier-Cabot, 1988), gradient (e.g. Mühlhaus and Aifantis, 1991), and Cosserat micro-polar (e.g. Mühlhaus and Vardoulakis, 1987) models. Alternatively, cohesive-crack models have been proposed under the assumption that damage can be represented by a dominant macro-fracture lumping all non-linearities into a discrete line (e.g. Hillerborg et al., 1976 and Bažant and Oh, 1983). That is, a fictitious crack concept is employed to represent the effect of a fracture process zone (FPZ) ahead of the crack tip, whereby phenomena such as small-scale yielding, micro-cracking or void growth and coalescence are assumed to take place. For the case of heterogeneous rocks, strain localization has also been successfully simulated by damage models with statistically distributed defects. A number of variations of this approach have been developed for numerical schemes such as FEM (Tang, 1997), FDM (Fang and Harrison, 2002), smooth-particle hydrodynamics (SPH) (Ma et al., 2011), cellular automaton (Feng et al., 2006), and lattice models (Blair and Cook, 1998).

Within continuum models, two approaches are commonly employed to account for the presence of rock mass discontinuities. If the number of discontinuities is relatively large, homogenization techniques are typically adopted. The most widely used homogenization approach consists of reducing, within a conventional elasto-plastic model, the rock mass deformation modulus and strength parameters to account for the degrading effect induced by the local geological conditions (Hoek et al., 2002 and Hoek and Diederichs, 2006). More advanced models can also include transversely isotropic elastic response induced by preferably oriented joints (Amadei, 1996) or failure-induced plastic anisotropic behavior (e.g. Mühlhaus, 1993 and Dyszlewicz, 2004). However, the classic homogenization approach is typically limited by the fact that slip, rotations and separation as well as size effects induced by discontinuities cannot be explicitly captured (Hammah et al., 2008). Alternatively, if the problem is controlled by a relatively low number of discrete features, special interface (or joint) elements can be incorporated into the continuum formulation (e.g. Goodman et al., 1968, Ghaboussi et al., 1973, Wilson, 1977, Pande and Sharma, 1979 and Bfer, 1985). This technique, also known as the combined continuum-interface method (Riahi et al., 2010), can accommodate large displacements, strains and rotations of discrete bodies. However, it is accurate as long as changes in edge-to-edge contacts along the interface elements are negligible throughout the solution (Hammah et al., 2007). That is, owing to the fixed interconnectivity between solid and joints and the lack of an automatic scheme to recognize new contacts, only small displacement/rotations along joints can be correctly captured (Cundall and Hart, 1992).

Discrete (or discontinuous) modeling techniques, commonly referred to as the discrete element method (DEM), treat the material directly as an assembly of separate blocks or particles. According to the original definition proposed by Cundall and Hart (1992), a DEM is any modeling technique that (i) allows finite displacements and rotations of discrete bodies, including complete detachment; and (ii) recognizes new contacts automatically as the simulation progresses. DEMs were originally developed to efficiently treat solids characterized by pre-existing discontinuities having spacing comparable to the scale of interest of the problem under analysis and for which the continuum approach described above may not provide the most appropriate computational framework. These problems include: blocky rock masses, ice plates, masonry structures, and flow of granular materials. DEMs can be further classified according to several criteria regarding, for instance, the type of contact between bodies, the representation of deformability of solid bodies, the methodology for detection and revision of contacts, and the solution procedure for the equations of motion (Jing and Stephansson, 2007). Based on the adopted solution algorithm, DEM implementations are broadly divided into explicit and implicit methods. The term *distinct element method* refers to a particular class of DEMs that use an explicit time-domain integration scheme to solve the equations of motion for rigid or deformable discrete bodies with deformable contacts (Cundall and Strack, 1979a). The most notable implementations of this group are arguably represented by the universal distinct element code (UDEC) (Itasca Consulting Group Inc., 2013) and the particle flow code (PFC) (Itasca Consulting Group Inc., 2012b). On the other hand, the best known implicit DEM is the discontinuous deformation analysis (DDA) method (Shi and Goodman, 1988). Despite the fact that DEMs were originally developed to model jointed structures and granular materials, their application was subsequently extended to the case of systems where the mechanical behavior is controlled by discontinuities that emerge as natural outcome of the deformation process, such as fracturing of brittle materials. Specifically, the introduction of bonding between discrete elements allowed capturing the formation of new fractures and, thus, extended the application of DEMs to simulate also the transition from continuum to discontinuum.

As observed by Bićanić (2003), the original boundary between continuum and discontinuum techniques has become less clear as several continuum techniques are capable of dealing with emergent discontinuities associated with the brittle fracture process. In particular, the hybrid approach known as the combined finite–discrete element method (FDEM) (Munjiza et al., 1995 and Munjiza, 2004) effectively starts from a continuum representation of the domain by finite elements and allows a progressive transition from a continuum to a discontinuum with insertion of new discontinuities.

The goal of this review paper is to provide a summary of selected discrete element and hybrid finite–discrete element modeling techniques that have recently emerged in the field of rock mechanics as simulation tools for fracturing processes in rocks and rock masses. Specifically, the commercially available codes PFC (Itasca Consulting Group Inc., 2012b), UDEC (Itasca Consulting Group Inc., 2013) and ELFEN (Rockfield Software Ltd., 2004) as well as the open-source software Yade (Kozicki and Donzé, 2008) and Y-Geo (Mahabadi et al., 2012a) are considered. Also, extensions of the DDA method to simulate fracturing processes are described. For each code, the fundamental implementation principles are illustrated with particular emphasis on the approach specifically adopted to simulate fracture nucleation and propagation and to account for the presence of rock mass discontinuities. The description of the governing principles is accompanied by a brief review of application studies focusing on laboratory-scale models of rock failure processes and on the simulation of damage development around underground excavations. For more extensive reviews of numerical methods in rock mechanics, the reader can refer to the work of Jing and Hudson (2002) and Jing (2003), with a detailed illustration of fundamentals and applications of DEMs provided by Jing and Stephansson (2007) and Bobet et al. (2009). Also, a review of modeling techniques for the progressive mechanical breakdown of heterogeneous rocks and associated fluid flow can be found in Yuan and Harrison (2006).

DISCRETE ELEMENT METHODS

Particle-based models: the PFC and Yade
Fundamental principles

Particle-based models were originally developed to simulate the micromechanical behavior of non-cohesive media, such as soils and sands (Cundall and Strack, 1979a). With this approach, the granular micro-structure of the material is modeled as a statistically generated assembly of rigid circular particles of varying diameters. The contacts between particles are typically assigned normal and shear stiffnesses as well as a friction coefficient. The commercially available code PFC represents an evolution of previous particle-based codes, namely BALL and TRUBAL (Cundall and Strack, 1979b), which applies cohesive bonds between particles to simulate the behaviors of solid rocks. The resultant model is commonly referred to as the bonded-particle model (BPM) for rock (Potyondy and Cundall, 2004). In a BPM, crack nucleation is simulated through breaking of internal bonds while fracture propagation is obtained by coalescence of multiple bond breakages. Blocks of arbitrary shapes can form as a result of the simulated fracturing process and can subsequently interact with each other.

Two types of bonds are typically used in PFC: the contact bond and the parallel bond. In the contact bond model, an elastic spring with constant normal and shear stiffnesses, k_n and k_s, acts at the contact points between particles, thus allowing only forces to be transmitted. In the parallel bond model, the moment induced by particle rotation is resisted by a set of elastic springs uniformly distributed over a finite-sized section lying on the contact plane and centered at the contact point (Fig. 1). This bond model reproduces the physical behavior of a cement-like substance gluing adjacent particles together.

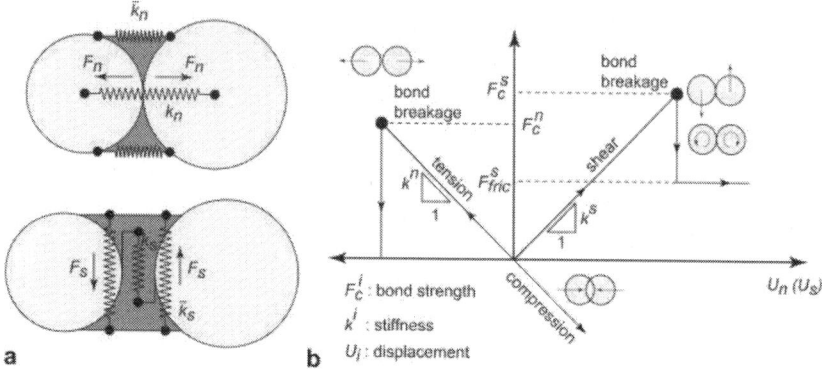

Figure 1. The parallel bond model implemented in PFC. (a) Normal and shear stiffnesses between particles. The contact stiffnesses, k_n and k_s, remain active even after the bond breaks as long as particles stay in contact. The bond stiffnesses (per unit area), $\bar{k}_n$ and $\bar{k}_s$, are suddenly removed when the bond breaks regardless of whether particles stay in contact or not. (b) Constitutive behavior in shear and tension (i = s, n). Figures redrawn after Potyondy and Cundall (2004) and Cho et al. (2007).

As further described in the next section, parallel bond rock models have been widely used to study fracturing and fragmentation processes in brittle rocks. However, one of the major drawbacks of this type of model is the unrealistically low ratios of the simulated unconfined compressive strength to the indirect tensile strength for synthetic rock specimens (Cho et al., 2007 and Kazerani and Zhao, 2010); the straightforward adoption of circular (or spherical) particles cannot fully capture the behavior of complex-shaped and highly interlocked grain structures that are typical of hard rocks. Furthermore, low emergent friction values are simulated in response to the application of confining pressure. To overcome these limitations, a number of enhancements to PFC were proposed. Potyondy and Cundall (2004) showed that by clustering particles together (Fig. 2a) more realistic macroscopic friction values can be obtained. Specifically, the intra-cluster bond strength is assigned a different strength value than the bond strength at the cluster boundary. Cho et al. (2007) showed that by applying a clumped-particle geometry (Fig. 2b) a significant reduction of the aforementioned deficiencies can be obtained, thereby allowing one to reproduce correct strength ratios, non-linear behavior of strength envelopes and friction coefficients comparable with laboratory values. More recently, Potyondy (2012) developed a new contact formulation, namely the flat-joint model, aimed at capturing the same effects of a clumped BPM (or of a grain-based UDEC model as described below) with a computationally more efficient method (Fig. 2c). The partial interface damage and continued moment-resisting ability of the flat-joint model allow the user to correctly match both the direct tensile and the unconfined compressive strengths of a hard rock.

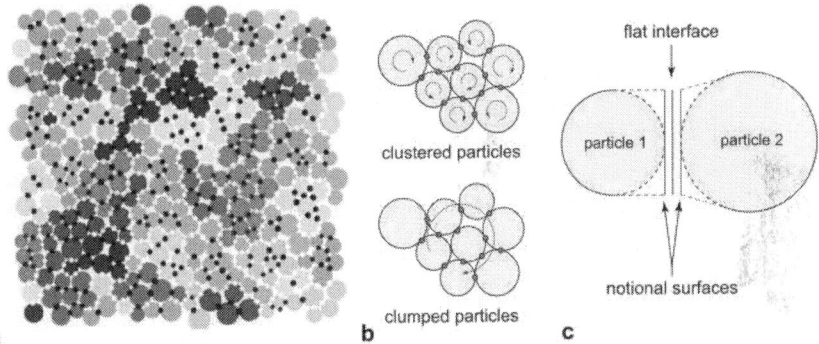

Figure 2. Proposed enhancements to the original BPM to capture realistic values of the ratio of unconfined compressive strength to indirect tensile strength. (a) Particle clustering (after Potyondy and Cundall (2004)), (b) clustered particles vs. clumped particles (after Cho et al. (2007), redrawn), and (c) flat-joint contact model showing the effective interface geometry (after Potyondy (2012), redrawn).

Another issue arising from the particle-based material representation of PFC is the inherent roughness of interface surfaces representing rock discontinuities (Fig. 3a). This roughness typically results in an artificial additional strength along frictional or bonded rock joints. This shortcoming was overcome by the development of the smooth-joint contact model (SJM) (Mas Ivars et al., 2008), which allows one to simulate a smooth interface regardless of the local particle topology (Fig. 3b). The combination of the BPM to capture the behavior of intact material with the SJM for joint network leads to the development of the so-called *synthetic rock mass* (Mas Ivars et al., 2011), which aims at numerically predicting rock mass properties, including scale effects, anisotropy and brittleness, that cannot be obtained using empirical methods.

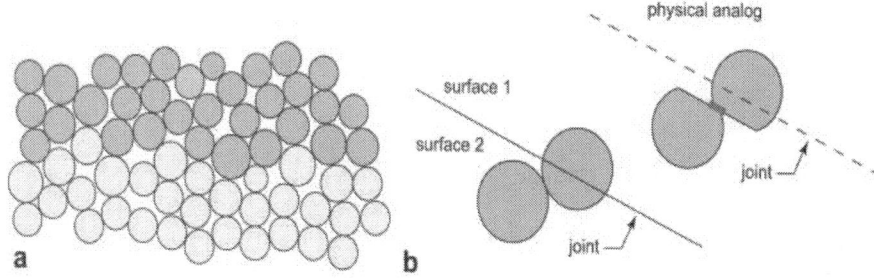

Figure 3. Representation of rock joints in PFC. (a) Traditional representation with rough surface, and (b) smooth-joint contact model. Figures redrawn after Mas Ivars et al. (2011).

The particle-based code Yade (Kozicki and Donzé, 2008, Kozicki and Donzé, 2009 and Šmilauer et al., 2010) has been recently introduced as an alternative modeling platform to the commercial software PFC described above. The major aims of the Yade project are (i) to provide enhanced flexibility in terms of adding new modeling capabilities and (ii) to promote code improvement through open-source development and direct feedback from the scientific community. In its basic formulation, the contact laws implemented in Yade share the same principles of those available in PFC. Small deformations are captured by linear elastic interaction forces between contacting discs/spheres. Rock fracturing is captured by the rupture of bonds, whose strength is characterized by a constant maximum acceptable force in tension and a cohesive-frictional maximum acceptable force in shear. Similar to PFC, the shear strength drops instantaneously to a purely frictional resistance after failure. Conversely, in tension, after the maximum force is reached, the stiffness can be varied by a softening factor, ζ, controlling the energy released due to bond breakage (Fig. 4a). Rock discontinuities can be treated in Yade using a contact logic analogous to the SJM of PFC (Scholtès and Donzé, 2012) (Fig. 3b). Specifically, the interactions between bonds crossing a prescribed discontinuity plane are identified and then re-oriented according to the joint surface, thus ensuring a frictional behavior that is independent of the inherent roughness induced by the particle topology. Furthermore, similar to the aforementioned SRM approach, sets of pre-existing discontinuities can be explicitly introduced in a Yade model as three-dimensional (3D) discrete fracture networks (Scholtès and Donzé, 2012). Applications of Yade to the investigation of the fundamentals of brittle rock failure have led to the implementation of an interaction range coefficient, γ, which can be used to link particles not directly in contact one with the other, yet located in the neighboring region (Scholtès and Donzé, 2013) (Fig. 4b). By doing so, the degree of interlocking between particles can be effectively controlled, thus allowing one to accurately model high ratios of compressive to tensile strengths as well as non-linear failure envelopes. This approach represents an alternative to the clumping logic and the flat-joint contact model of PFC.

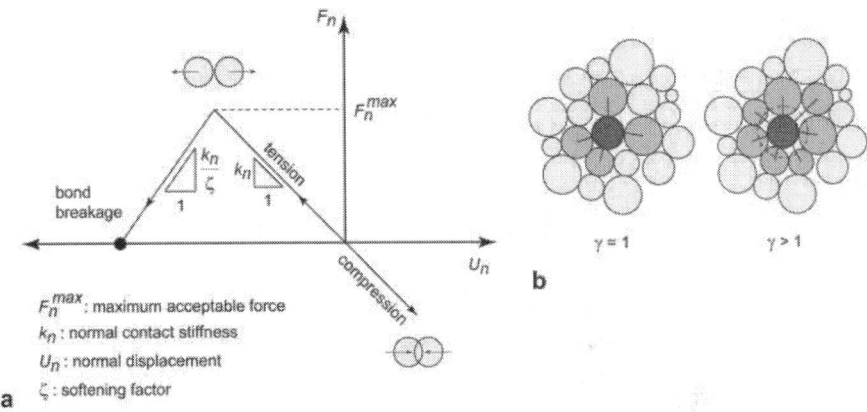

Figure 4. The particle-based code Yade. (a) Interaction law between particle in tension and compression, and (b) effect of the interaction range coefficient, γ, on the simulated contact fabric. Figures redrawn afterScholtès and Donzé (2013).

Main advantages of the particle-based modeling methodology include the simple mathematical treatment of the problem, whereby complex constitutive relationships are replaced by simple particle contact logic, and the natural predisposition of the approach to account for material heterogeneity. On the other hand, considering the high level of simplification introduced, extensive experimental validation is needed to verify that the method can capture the observed macroscopic behavior of rock. Moreover, an extensive calibration based on experimentally measured macro-scale properties is required to determine the contact parameters that will predict the observed macro-scale response.

Applications

PFC has been extensively used within the rock mechanics community to numerically investigate the fundamental processes of brittle fracturing in rocks by means of laboratory-scale models. Potyondy et al. (1996) first proposed a synthetic PFC model that could reproduce modulus, unconfined compressive stress, and crack initiation stress of the Lac du Bonnet Granite. Extended results were illustrated by Potyondy and Cundall (2004) with the simulation of the stress–strain behavior during biaxial compression tests for varying confining pressures. Several features of the rock behavior emerged from the BPM, including elasticity, fracturing, damage accumulation producing material anisotropy, dilation, post-peak softening and strength increase with confinement. Since PFC simulates quasi-static deformation by solving the equations of motion, elasto-dynamics effects, such as stress wave propagation and cracking-induced AE, can be explicitly simulated. In this context, Hazzard and Young (2000) developed a technique to dynamically quantify AE in a PFC model. The approach was validated by simulating the seismic b value of a confined test on granite. The aforementioned approach was further improved by introducing moment tensor calculation based on change in contact forces upon particle contact breakage and was applied to the micro-seismic simulation of a mine-by experiment in a crystalline rock (Hazzard and Young, 2002) and of an excavation-induced fault slip event (Hazzard et al., 2002). 3D simulations of acoustic activity using PFC3D were proposed by Hazzard and Young (2004). Diederichs (2003)used PFC simulations to explore the aspects of grain-scale tensile damage accumulation under both macroscopically tensile and compressive conditions. A BPM was employed as numerical analog to study the effects of tensile damage and the sensitivity to low confinement in controlling the failure of hard rock masses in proximity of underground excavations. Application of PFC to determining the fracture toughness of synthetic rock-like specimen was illustrated by Moon et al. (2007). Analyses of failure and deformation mechanisms during direct shear loading of rock joints have also been carried out to obtain original insights into rock fracture shear behavior and asperity degradation.Rasouli and Harrison (2010) investigated the relation between Riemannian roughness parameter and shear strength of profiles comprising symmetric triangular asperities sheared at different normal stress levels. Asadi et al. (2012) extended the previous results with consideration of the shear strength and asperity degradation processes of several synthetic profiles including triangular, sinusoidal and randomly generated profiles. Zhao (2013) simulated single- and multi-gouge particles in a rough fracture segment undergoing shear and analyzed the behavior of gouge particles as function of the applied confinement. Another important mechanism of the failure process in rocks, such as the initiation and propagation of cracks from pre-existing flaws, has been analyzed using BPM. Zhang and Wong (2012) numerically simulated the cracking process in rock-like material containing a single flaw under uniaxial compression, whileZhang and Wong (2013) investigated the coalescence behavior for the case of two stepped and coplanar pre-existing open flaws. The effect of confinement on wing crack propagation was studied by Manouchehrian and Marji (2012).

BPMs have been successfully applied to the study of damaged zones around underground openings. The spalling phenomena observed around the Atomic Energy of Canada Limited's (AECL) mine-by experiment tunnel (Martin et al., 1997) were first simulated by Potyondy and Cundall (1998). Further analysis of the notch formation process in terms of coalescence of ruptured bonds was provided by Potyondy and Cundall (2004) using a PFC2D model embedded in a continuum finite-difference model (Fig. 5a). Hazzard and Young (2002) provided a micro-seismic simulation of the same excavation by comparing the actual seismicity recorded underground with the simulated spatial and temporal distribution of events. The effect of low stiffness spray-on liner of fracture propagation based on *in situ* conditions of the above mentioned mine-by experiment was numerically studied by Tannant and Wang (2004). Similarly, Potyondy and Cundall (2000) used PFC2D to predict damage formation adjacent to a circular excavation in an anisotropic gneissic tonalite at the Olkiluoto deep geological repository.Fakhimi et al. (2002) showed that a BPM could match failure load, crack pattern, and spalling observed during a biaxial compression test on a sandstone specimen with a circular opening (Fig. 5b). Numerical studies on thermally-induced fracturing around openings in granite were carried out by Wanne and Young (2008) and Wanne (2009) for a laboratory-scale heater experiment and the AECL's Tunnel Sealing Experiment, respectively. Sagong et al. (2011) investigated the influence of the joint angle on the rock fracture and joint sliding behaviors around an opening in a jointed rock model.

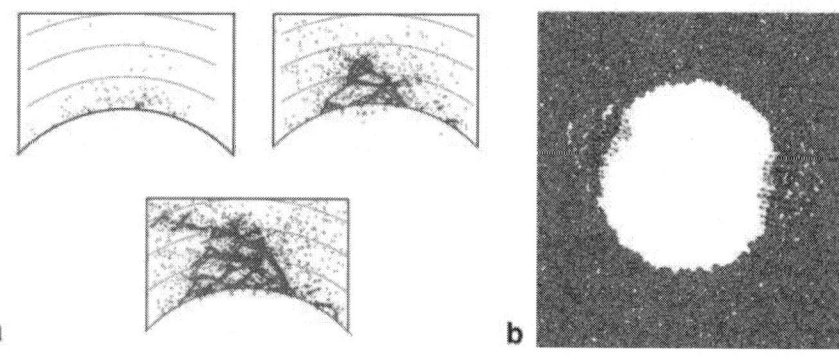

Figure 5. Simulation of fracture development around underground excavation using PFC. (a) Modeling of notch formation around the AECL's mine-by experiment tunnel (after Potyondy and Cundall (2004)). (b) Damaged notches around a hole under biaxial compression (after Fakhimi et al. (2002)).

Simulation studies using the open-source code Yade have been focused to date on the role played by discrete fracture networks (DFNs) (Scholtès et al., 2011, Harthong et al., 2012 and Scholtès and Donzé, 2012) and grain interlocking (Scholtès and Donzé, 2013) in controlling the mechanical responses of 3D rock samples.

The universal distinct element code (UDEC)
Fundamental principles

In UDEC the computational domain is discretized into blocks using a finite number of intersecting discontinuities. Each block is internally subdivided using a finite difference, or a finite volume, scheme for calculation of stress, strain and displacement. Model deformability is captured by an explicit, large strain Lagrangian formulation similar to the continuum code FLAC (Itasca Consulting Group Inc., 2012a). The mechanical interaction between blocks is characterized by compliant contacts using a finite stiffness together with a tensile strength criterion in the normal direction, and a tangential stiffness together with a shear strength criterion (e.g. Coulomb-type friction) in the tangential direction to the discontinuity surface. Similar to PFC, static problems are treated using a dynamic relaxation technique by introducing viscous damping to achieve steady state solutions.

When using the classic formulation of UDEC, rock failure is captured either in terms of plastic yielding (e.g. Mohr–Coulomb criterion with tension cut-off) of the rock matrix or displacements (i.e. sliding, opening) of the pre-existing discontinuities. That is, new discontinuities cannot be driven within the continuum portion of the model and therefore discrete fracturing through intact rock cannot be simulated. However, Lorig and Cundall (1989) showed that this shortcoming can be overcome by introducing a polygonal block pattern, such as the Voronoi tessellation, to the UDEC capability. As depicted in Fig. 6, a physical discontinuity is created when the stress level at the interface between block exceeds a threshold value either in tension or in shear. Although new fractures are so propagated, this technique is not based on a linear elastic fracture mechanics (LEFM) approach. That is, unlike classic LEFM models, fracture toughness and stress intensity factors are not considered. Furthermore, material softening in the FPZ, typically captured using cohesive-crack models, is disregarded.

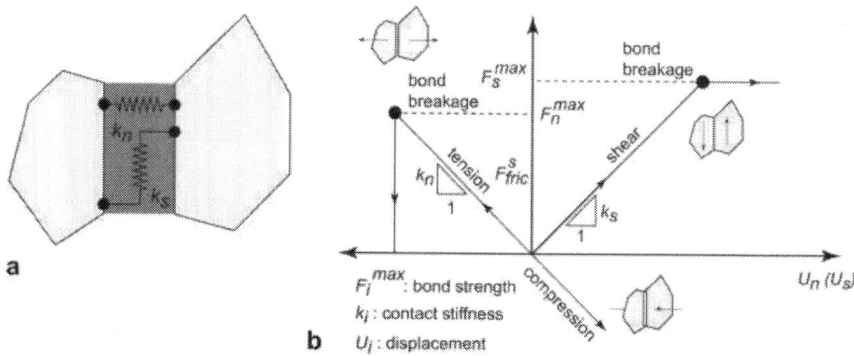

Figure 6. UDEC modeling of fracture propagation in rock. (a) Normal and shear stiffnesses between blocks. (b) Constitutive behavior in shear and tension (i = s, n). Figures redrawn after Kazerani and Zhao (2010).

Although polygonal block models are computationally more expensive than particle-based ones, they can provide a more realistic representation of the rock micro-structure (Lemos, 2012). Owing to the full contact between grains and better interlocking offered by the random polygonal shapes, the grain-based UDEC model overcomes some of the limitations of parallel-bonded particle models, as further described below.

Applications
Owing to the above mentioned characteristics, grain-based DEMs have been employed to study the fracturing behavior of rocks. Christianson et al. (2006) used a grain-based UDEC model to numerically complement laboratory testing on a lithophysal rock under confined conditions. The mechanical degradation of the same rock type was investigated by Damjanac et al. (2007) using a similar technique. The model was then upscaled to study the stability of emplacement drifts at Yucca Mountain under mechanical, thermal, and seismic loading as well as time-dependent effects. Using a UDEC-Voronoi model, Yan (2008) investigated the laboratory-scale step-path failure (e.g. wing cracking and fracture coalescence) in a sample containing pre-existing joints with application to slope stability problems. A similar approach was adopted by Lan et al. (2010) to numerically assess the effect of heterogeneity on the

micromechanical extensile behavior during compressive loading on Lac du Bonnet Granite and Äspö Diorite. The model directly incorporated several sources of heterogeneity, including microgeometric heterogeneity, grain-scale elastic heterogeneity, and microcontact heterogeneity. A calibration procedure to determine a unique set of micro-parameters for a grain-based UDEC model was developed by Kazerani and Zhao (2010) (Fig. 7a). A series of numerical experiments (i.e. uniaxial/biaxial compression and Brazilian tension) were used to assess the relationship between macro- and micro-parameters. The model was also shown to correctly capture the ratio of compressive to tensile strengths of rock samples measured in the laboratory, therefore overcoming some of the original defects of parallel-bonded particle models. Finally, the grain-based UDEC approach was also employed to study the effect of joint persistence on the evolution of damage during direct shear tests (Alzo'ubi, 2012).

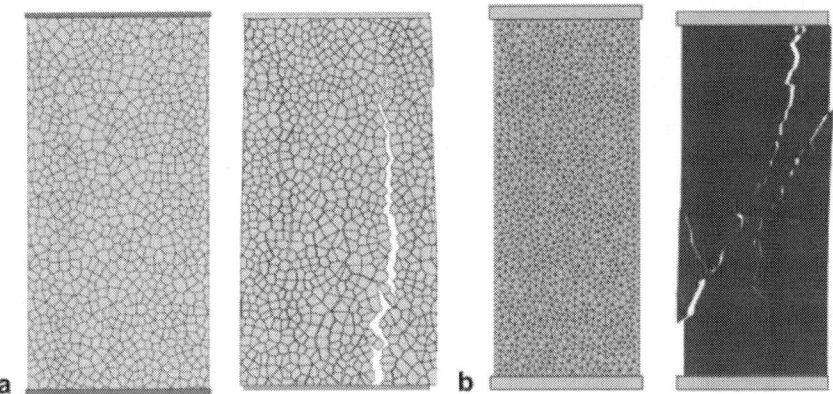

Figure 7. Simulations of rock failure under compression using UDEC. (a) Uniaxial compression test on Augite granite using a grain-based model (after Kazerani and Zhao (2010)). (b) Uniaxial compression test on plaster using a cohesive boundary approach (after Kazerani et al. (2012)).

Last, it is worth mentioning three approaches that were proposed to capture fracture processes in UDEC as an alternative to the adoption of a polygonal structure. Firstly, based on fracture mechanics considerations, a time-dependent joint cohesion was implemented by Kemeny (2005) to capture the progressive mechanical degradation during the failure of rock bridges along discontinuities. The model was validated using several laboratory-scale examples and then was used to investigate the time-dependent degradation of drifts for the storage of nuclear waste at Yucca Mountain. Secondly, Jiang et al. (2009) developed an expanded distinct element method (EDEM) based on UDEC whereby potential cracks, with bonding strengths equivalent to the rock matrix, are pre-distributed within the model based on the plastic regions and direction of principal stresses obtained from a preliminary elasto-plastic analysis. The approach was applied to the simulation of cracking around a large underground excavation in a blocky rock mass. Lastly, Kazerani et al. (2012) implemented a UDEC model which represents the rock material as a collection of irregular-sized deformable triangles with cohesive boundaries controlling the material fracture and fragmentation properties (Fig. 7b). A reasonable agreement was found between numerical simulation and experimental laboratory results of compressive, tensile and shear tests on plaster (Kazerani, 2013). As further illustrated in Section 3.2, this model shares several characteristics of the Y-Geo implementation of the combined FDEM.

The discontinuous deformation analysis (DDA) method

The DDA method is an implicit DEM originally proposed by Shi and Goodman (1988) to simulate the dynamics, kinematics, and elastic deformability of a system contacting rock blocks. Similar to classic finite element formulations, the governing equations are represented by a global system of linear equations which are obtained by minimizing the total potential energy of the system. Displacements and strains are taken as variables and the stiffness matrix of the model is assembled by differentiating several energy contributions including block strain energies, contacts between blocks, displacement constraints and external loads. An implicit formulation is used to solve the system of equations. In the basic DDA implementation, each block is simply deformable as the strain and stress fields are constant over the entire block area. However, improved deformability models can be achieved by introducing higher order strain fields or by subdividing each block into a set of simply deformable sub-blocks (Lin et al., 1996). The imposition of contact constraints between blocks can be obtained by a number of methods including the penalty method, the Lagrange multiplier method or the augmented Lagrangian method. The frictional behavior along block interfaces is modeled by a Mohr–Coulomb criterion.

Traditionally, DDA simulations have been employed to capture failure along predefined structural planes in blocky rock masses (e.g. Hatzor and Benary, 1998, Bakun-Mazor et al., 2009 and Hatzor et al., 2010). Nevertheless, some attempts have been made to introduce fracturing capabilities within the DDA framework. The simplest technique is similar to the UDEC-Voronoi approach: fracturing is captured as debonding of artificial block interfaces if a Mohr–Coulomb with tension cut-off criterion is locally violated (Ke, 1997). A second approach involves comparing the maximum and minimum principal stresses calculated at each block centroid with a three-parameter Mohr–Coulomb criterion. If the criterion is satisfied, appropriate fracture plane directions are computed (i.e. one axial splitting plane in tension or two conjugated planes in shear) and the block is then subdivided into multiple sub-blocks. Simple validation of this method was presented by Koo and Chern (1997) using uniaxial and confined compression as well as uniaxial tension test models. Finally, Lin et al. (1996) proposed an algorithm based on an iterative sub-block approach which allows one to capture continuous crack growth from the tip of a pre-existing flaw.

Recently, a new method, known as the discontinuous deformation and displacement (DDD) method, has been developed by Tang and Lü (2013) by merging DDA with a continuum-based damage mechanics code, namely the rock failure process analysis (RFPA) program (Tang, 1997). The model aims at combining the advantages of DDA to capture the large-displacement mechanics of discontinuous systems with the capabilities of RFPA to simulate the small-strain deformational mechanisms characterizing the failure process of intact rock material.

THE HYBRID FINITE–DISCRETE ELEMENT METHOD (FDEM)

In the hybrid continuum–discontinuum technique known as the combined FDEM, the simulation starts with a continuous representation of the solid domain of interest. As the simulation progresses, typically through explicit integration of the equations of motion, new discontinuities are allowed to form upon satisfying some fracture criterion, thus leading to the formation of new discrete bodies. In general, the approach blends FEM techniques with DEM concepts (Barla and Beer, 2012). The latter algorithms include techniques for detecting new contacts and for dealing with the interaction between discrete bodies, while the former techniques are used for the computation of internal forces and for the evaluation of a failure criterion and the creation of new cracks. Hybrid finite–discrete element models should not be confused with coupled continuum–discontinuum approaches (e.g. Pan and Reed, 1991 and Billaux et al., 2004) which represent the problem of far- and near-fields using continuum-based and DEM techniques, respectively.

In the following sections, the fundamental principles of two common FDEM implementations are briefly illustrated, namely ELFEN (Rockfield Software Ltd., 2004) and Y-Geo (Munjiza, 2004 and Mahabadi, 2012), as well as their applications to the study of the fracturing behavior of rocks.

ELFEN

Fundamental principles

The continuum formulation of ELFEN is based on the explicit finite element method. Material softening (or hardening) is captured using a non-associative Mohr–Coulomb elasto-plastic model with shear strength parameters, including cohesion, friction angle and dilation, defined as function of the effective plastic strain. The localization of strain is obtained by regularizing the standard description of the continuum with the incorporation of fracture mechanics principles in the equations governing the evolution of state variables (Owen and Feng, 2001 and Klerck et al., 2004). In particular, material softening associated with fracturing is captured under the main assumption that the quasi-brittle failure is extensional in nature. As thoroughly described by Klerck (2000), the extensional failure is modeled directly and indirectly for the cases of tensile stress and compressive stress fields, respectively. Under direct tension, several constitutive models can be used such as the rotating crack and the Rankine tensile smeared crack. With these models, material strain softening is fully governed by the tensile strength and the specific fracture energy parameters. Under compressive stress fields, a Mohr–Coulomb yield criterion is combined with a fully anisotropic tensile smeared crack model. With this approach, known as the *compressive fracture model*, the extensional inelastic strain associated with the dilation response is explicitly coupled with the tensile strength in the dilation direction. That is, increments of extensional strain are associated with tensile strength degradation in the perpendicular direction.

Upon localization of damage into crack bands and complete dissipation of the fracture energy, a discrete fracture is realized. Hence, the transition from continuous to discontinuous behavior involves transferring a virtual smeared crack into a physical discontinuity in the finite element mesh (Owen and Feng, 2001). The mesh topology update is based on a nodal fracture scheme with all new fractures developing in tension (i.e. Mode I) in the direction orthogonal to the principal stress direction where the tensile strength becomes zero. This procedure is numerically accomplished by first creating a non-local failure map for the whole domain based on the weighted nodal averages of a failure factor, defined as the ratio of the inelastic fracturing strain to the critical fracturing strain. Secondly, a failure direction is determined for each nodal point where the failure factor is greater than one based on the weighted average of the maximum failure strain directions of all elements connected to the node. Finally, a discrete crack is inserted through the failure plane. As depicted in Fig. 8, the insertion of a new crack can be accomplished using two different algorithms (Klerck et al., 2004). The *intra-element* insertion drives a new fracture along the crack propagation direction by directly splitting the finite elements. In this case, a local adaptive re-meshing may be necessary to achieve an acceptable element topology and avoid highly-skewed sliver elements that could decrease the numerical stability threshold of the integration time step. Conversely, with the *inter-element* insertion, the discrete crack is snapped to the existing element edge most favorably oriented with respect to the failure plane. Following the crack insertion, the damage variables in the adjacent finite elements are set to zero and the contact along the two newly-created surfaces is treated using a contact interaction algorithm (e.g. penalty or Lagrangian multiplier method).

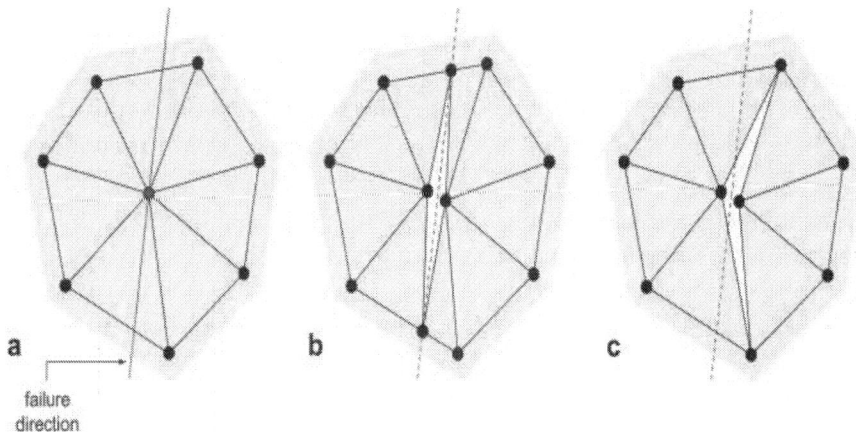

Figure 8. Nodal fracture scheme of ELFEN. (a) Initial state before fracturing, (b) intra-element crack insertion, and (c) inter-element crack insertion. Figure redrawn after Klerck (2000).

Applications

First applications of ELFEN to modeling rock failure under compressive loads were proposed by Klerck (2000). In particular, the simulation of borehole breakouts indicated that the aforementioned compressive fracture model of ELFEN can describe both the axial splitting typical of brittle materials and the shear failure usually observed in more ductile ones (Fig. 9). A study of fracture development around deep tunnels was proposed by Sellers and Klerck (2000) using laboratory-scale models. As observed in the laboratory, the model could reproduce fractures developing sub-parallel to the excavation boundary as well as the confining effect of pre-existing joints on the development of damage. Moreover, the experimentally measured acoustic activity was analyzed using the release of kinetic energy simulated by the model as numerical equivalent. Coggan et al. (2003) described several examples of rock engineering application of ELFEN, including stability analysis of roof beams and pillars in underground excavations and rock slope stability problems. Klerck et al. (2004)proposed a quantitative analysis of compression tests on rocks with direct comparison of the load-displacement response and of the evolution of fracture development with experimental data (Fig. 10a). This type of analysis was subsequently extended to 3D models that aimed at studying the influence of the intermediate principal stress on rock fracturing and strength near excavation boundaries (Cai, 2008). The results clearly showed that the generation of tunnel surface parallel fractures and microcracks can be attributed to material heterogeneity and to the existence of relatively high intermediate principal stress as well as zero to low minimum principal stress confinement (Fig. 10b). Investigations of failure behavior of rock specimens under indirect tensile stress conditions were first proposed by Cai and Kaiser (2004) and, more recently, by Cai (2013). The simulation of crack initiation and propagation from a pre-existing flaw highlighted the influence of the flaw frictional resistance on the development of primary and secondary cracks as well as on the failure load. Application of ELFEN to the investigation of damage mechanisms (e.g. surface wear and tensile fracturing) along joint planes under direct shear conditions was illustrated by Karami and Stead (2008). A discrete fracture rock mass model was proposed by Pine et al. (2006) by combining an ELFEN model with the fracture geometry generated by a DFN software. The approach was used to obtain insights into the influence of pre-existing joints on the rock mass behavior in underground pillars, rock slides, and block caving operations (Eberhardt et al., 2004, Pine et al., 2007, Elmo and Stead, 2010, Vyazmensky et al., 2010a and Vyazmensky et al., 2010b). Yan (2008) numerically analyzed fracture coalescence and rock failure mechanisms in laboratory-scale specimens containing step-path fractures. The study was also extended to the simulation of rock bridge fracture associated with potential toe breakout failure in large open pit slopes. Original application of ELFEN to the investigation of failure behavior of layered rocks can be found inStefanizzi (2007) and Stefanizzi et al. (2008).

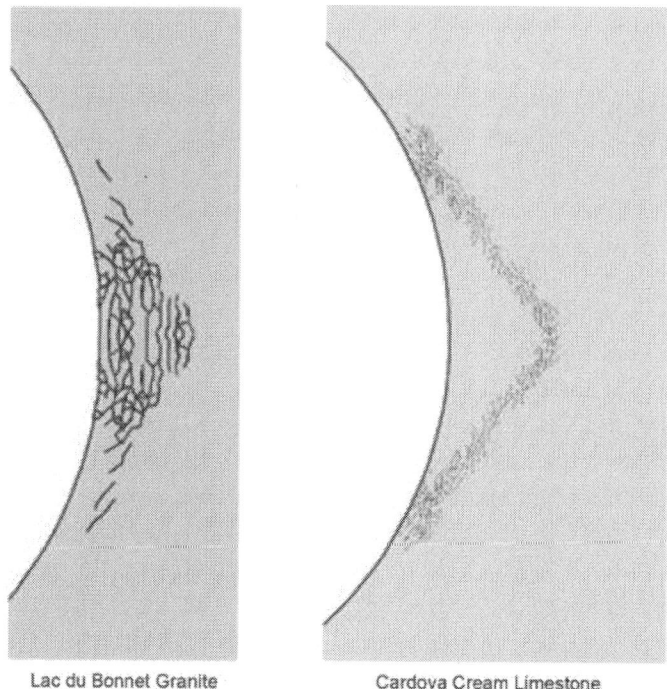

Lac du Bonnet Granite   Cardova Cream Limestone

Figure 9. Simulation of borehole breakout in different rock types using ELFEN; left: brittle failure, right: ductile failure (after Klerck (2000)).

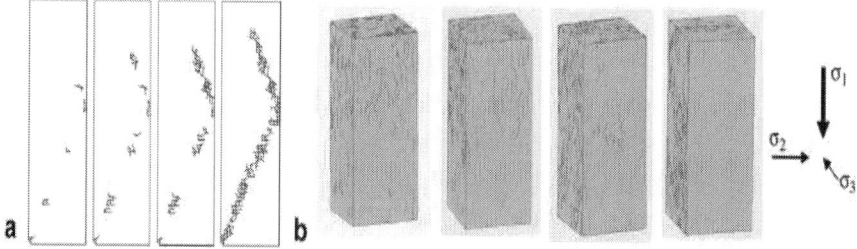

Figure 10. ELFEN simulation of rock failure under compression. (a) Evolution of fracturing during the simulation of a confined compression test on sandstone (after Klerck et al. (2004)). (b) Final fracture patterns for triaxial compression test simulations for increasing values of the intermediate principal stress, σ_2 (after Cai (2008)).

Y-Geo
Fundamental principles

The continuum representation of Y-Geo is based on the discretization of the modeling domain with three-noded triangular elements together with four-noded cohesive elements embedded between the edges of all adjacent triangle pairs (Fig. 11a). The elastic deformation of the bulk material is captured by the constant-strain, linear-elastic triangular elements with impenetrability between elements enforced by a penalty function method (Munjiza and Andrews, 2000). Fracture nucleation within the continuum is simulated by the breakage of the cohesive elements (Munjiza et al., 1999). Since fractures can nucleate only along the boundaries of the triangular elements, arbitrary fracture trajectories can be reproduced within the constraints imposed by the initial mesh topology. Unlike ELFEN, the mesh topology in Y-Geo is never updated during the simulation and re-meshing is not performed. Consequently, a sufficiently small element size should be adopted to reproduce the correct mechanical response (Munjiza and John, 2002).

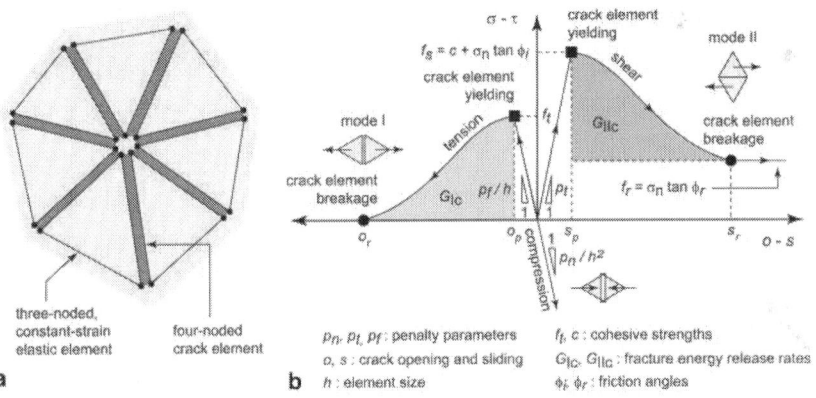

Figure 11. Simulation of fracture propagation with Y-Geo. (a) Representation of a continuum using cohesive elements interspersed throughout a mesh of triangular elements. (b) Constitutive behavior of the crack elements defined in terms of bonding stress (tensile, σ, and shear, τ) vs. relative displacement (opening, o, and sliding, s) between the edges of adjacent triangular elements.

The constitutive behavior of crack elements is implemented in terms of relative displacement between opposite triangle edges and incorporates principles of non-linear elastic fracture mechanics (Fig. 11b). Material starts to yield, in tension or shear, upon reaching a displacement value corresponding to the peak cohesive strengths. The Mode I peak strength is defined by a constant tensile strength, f_t, while the Mode II peak strength, f_s, is computed according to the Mohr–Coulomb criterion. The complete breakage of the crack element and thus the nucleation of a discrete crack are accomplished after dissipating the material fracture energy release rate, G_c. Upon breakage of the cohesive surface, the crack element is removed from the simulation and therefore the model transition from a continuum to discontinuum is locally completed. A Coulomb-type friction is applied along every newly-created fracture. As the simulation progresses, finite displacements and rotations of discrete bodies are allowed and new contacts are automatically recognized (Munjiza and Andrews, 1998).

Given the above modeling assumptions, the Y-Geo implementation of FDEM, rather than ELFEN's approach, can be considered closer to a fully discrete method. In ELFEN, a real transition from a continuous elasto-plastic medium to a solid with discrete fractures is accomplished by dynamically inserting cracks into the model. Conversely, the material representation of Y-Geo resembles one of a particle-based DEM with rigid particles and deformable particle bonds replaced by deformable triangles and cohesive elements, respectively.

Applications

Early applications of Y-Geo aimed at validating the adopted FDEM implementation in the context of the failure processes typically observed during standard rock mechanics tests on brittle rocks. Mahabadi et al. (2009) investigated the influence of loading rate and sample orientation during a Brazilian test simulation on a layered rock. Mahabadi et al. (2010b) simulated the behavior of a homogeneous rock sample under biaxial loading conditions. The model captured the main phenomena observed in a triaxial test including localization of failure, fracture initiation and evolution, increase of strength with confinement, and brittle–ductile transition. Mahabadi   et al.   (2010a) showed   good   agreement between the experimental results of a high-strain-rate Brazilian disc test and Y-Geo simulation results in terms of tensile strength, failure time, and fracture mechanisms. Subsequently, the numerical experimentation with Y-Geo focused on the mechanical behavior of heterogeneous granitic rocks under different loading conditions. A procedure to incorporate actual micromechanical input parameters of the rock, together with the real distribution of mineral constituents, into a FDEM model was developed by Mahabadi et al. (2012b). Overall, the simulation results showed that by including accurate micromechanical parameters and the intrinsic rock geometric features, such as spatial phase heterogeneity and microcracks, the model can more accurately predict the mechanical responses of rock specimens under indirect tension (Mahabadi, 2012). Moreover, heterogeneity was shown to play a key role in controlling the non-linear stress–strain behavior associated with the damage progress under compressive loading conditions (Fig. 12a). Further validation of the same model was also carried out by considering the acoustic activity reproduced during the numerical experiments (Lisjak et al., 2013).

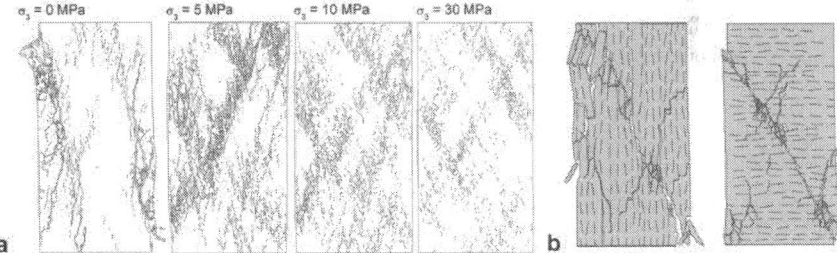

Figure 12. Simulation of rock fracture under compression using Y-Geo. (a) Final fracture patterns of a biaxial compression test simulation on a heterogeneous rock sample at increasing confining pressures, σ_3 (afterMahabadi et al. (2012b)). (b) Final fracture patterns of a unconfined compression test simulation on layered rock samples (after Lisjak et al. (2014a)).

More recently, the capabilities of Y-Geo were extended to capture the behavior of transversely isotropic rocks (Lisjak   et al., 2014, Lisjak   et al., 2013a and Lisjak, 2013). The methodology was validated by investigating the laboratory-scale fracturing behavior of a clay shale, namely Opalinus Clay (Fig. 12b), and the formation process of the excavation damaged zone (EDZ) around a test tunnel at the Mont Terri rock laboratory (Fig. 13). Finally, Rougier et al. (2011) developed a 3D version of the Y-code of Munjiza (2004) and analyzed the effect of energy dissipation during the simulation of split Hopkinson pressure bar Brazilian experiments.

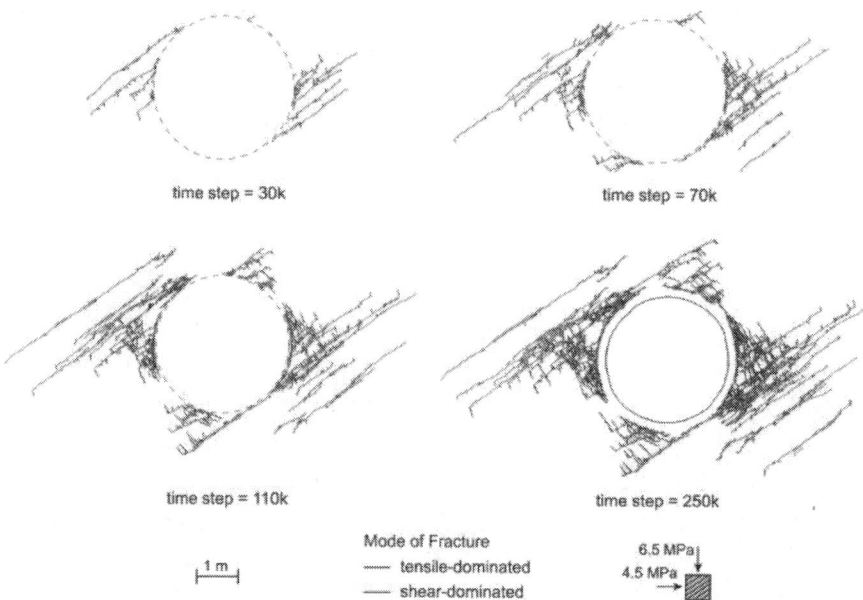

Figure 13. Simulation sequence of the EDZ formation process in a bedded rock using Y-Geo (after Lisjak (2013)).

CONCLUDING REMARKS

Modeling the failure process of brittle rocks represents a challenging task owing to several intrinsic features of the material mechanical behaviors, including strain localization, non-linear stress–strain response, and confinement dependent characteristics. At the rock mass scale, pre-existing discontinuities, such as faults and bedding planes, contribute to introducing further complexity into the problem. This review tried to summarize the use of both DEMs (either particle- or grain-based) and combined FDEM codes and their use to investigate the failure behavior of rocks under a variety of loading conditions. The introduction of bonds between discrete elements has extended the application of the DEM, which was originally developed to simulate the behavior of solids whereby pre-existing discontinuities have spacing comparable to the scale of interest of the problem under analysis (e.g. blocky rock masses, granular media), to capture the growth of new fractures. More recently, the FDEM, by combining fracture mechanics with FEM and DEM, has emerged as an appealing alternative numerical tool for rock mechanics applications where an explicit consideration of fracture and fragmentation processes is of paramount importance.

REFERENCES

1. Alzo'ubi AK. Modeling of rocks under direct shear loading by using discrete element method. Alhosn University Journal of Engineering & Applied Sciences 2012;4: 5e20.
2. Amadei B. Importance of anisotropy when estimating and measuring in situ stresses in rock. International Journal of Rock Mechanics and Mining Sciences & Geomechanics Abstracts 1996;33(3):293e325.
3. Asadi MS, Rasouli V, Barla G. A bonded particle model simulation of shear strength and asperity degradation for rough rock fractures. Rock Mechanics and Rock Engineering 2012;45(5):649e75.
4. Bakun-Mazor D, Hatzor YH, Dershowitz WS. Modeling mechanical layering effects on stability of underground openings in jointed sedimentary rocks. International Journal of Rock Mechanics and Mining Sciences 2009;46(2):262e71.
5. Barla M, Beer G. Special issue on advances in modeling rock engineering problems. International Journal of Geomechanics 2012;12:617. Bazant ZP, Oh BH. Crack band theory for fracture of concrete. RILEM Materials and Structures 1983;16(3):155e77.
6. Bazant ZP, Pijaudier-Cabot G. Nonlocal continuum damage, localization instability and convergence. Journal of Applied Mechanics 1988;55(2):287e93.

7. Benson PM, Vinciguerra S, Meredith PG, Young RP. Laboratory simulation of volcano seismicity. Science 2008;322(5899):249e52.

8. Bfer G. An isoparametric joint/interface element for finite element analysis. International Journal for Numerical Methods in Engineering 1985;21(4):585e600.

9. Bicanic N. Fragmentation and discrete element methods. In: Milne I, Ritchie RO, Karihaloo B, editors. Comprehensive Structural Integrity. Oxford: Pergamon; 2003. pp. 427e57.

10. Bieniawski ZT. Mechanism of brittle fracture of rock. Part I: theory of the fracture process. International Journal of Rock Mechanics and Mining Sciences & Geomechanics Abstracts 1967;4(4):405e6.

11. Billaux D, Dedecker F, Cundall P. A novel approach to studying rock damage: the three dimensional Adaptive Continuum/Discontinuum Code. In: Schubert W, editor. Proceedings of the ISRM Regional Symposium Eurock 2004 and the 53rd Geomechanics Colloquium. Salzburg: Austria; 2004.

12. Blair SC, Cook NGW. Analysis of compressive fracture in rock using statistical techniques: part I. A non-linear rule-based model. International Journal of Rock Mechanics and Mining Sciences 1998;35(7):837e48.

13. Bobet A, Fakhimi A, Johnson S, Morris J, Tonon F, Yeung MR. Numerical models in discontinuous media: review of advances for rock mechanics applications. Journal of Geotechnical and Geoenvironmental Engineering, ASCE 2009; 135(11):1547e61.

14. Brace WF, Paulding Jr BW, Scholz C. Dilatancy in the fracture of crystalline rocks. Journal of Geophysical Research 1966;71(16):3939e53.

15. Cai M, Kaiser PK. Numerical simulation of the Brazilian test and the tensile strength of anisotropic rocks and rocks with pre-existing cracks. International Journal of Rock Mechanics and Mining Sciences 2004;41(Suppl. 1):478e83.

16. Cai M. Fracture initiation and propagation in a Brazilian disc with a plane interface: a numerical study. Rock Mechanics and Rock Engineering 2013;46(2):289e302.

17. Cai M. Influence of intermediate principal stress on rock fracturing and strength near excavation boundariesdinsight from numerical modeling. International Journal of Rock Mechanics and Mining Sciences 2008;45(5):763e72.

18. Cho N, Martin CD, Sego DC. A clumped particle model for rock. International Journal of Rock Mechanics and Mining Sciences 2007;44(7):997e1010.

19. Christianson M, Board M, Rigby D. UDEC simulation of triaxial testing of lithophysal tuff. In: Proceedings of the 41st US Symposium on Rock Mechanics (USRMS). Golden, USA: American Rock Mechanics Association; 2006.

20. Coggan JS, Pine RJ, Stead D, Rance JM. Numerical modelling of brittle rock failure using a combined finite-discrete element approach: implications for rock engineering design. In: Technology Roadmap for Rock Mechanics: the 10th ISRM Congress. Gauteng, South Africa: South African Institution of Mining and Metallurgy; 2003. pp. 211e8.

21. Cundall PA, Hart RD. Numerical modelling of discontinua. Engineering Computations 1992;9:101e13.

22. Cundall PA, Strack ODL. A discrete numerical model for granular assemblies. Geotechnique 1979a;29(1):47e65.

23. Cundall PA, Strack ODL. The distinct element method as a tool for research in granular media e part II. Rep. NSF Grant ENG76-20771. Minneapolis, USA: Department of Civil & Mineral Engineering, University of Minnesota; 1979b.

24. Damjanac B, Board M, Lin M, Kicker D, Leem J. Mechanical degradation of emplacement drifts at Yucca Mountain e a modeling case study: part II: lithophysal rock. International Journal of Rock Mechanics and Mining Sciences 2007;44(3):368e99.

25. De Borst R, Crisfield MA, Remmers JJC, Verhoosel CV. Non-linear finite element analysis of solid and structures. 2nd ed. Chichester, UK: John Wiley & Sons Ltd.; 2012.

26. De Borst R, Sluys LJ, Mühlhaus HB, Pamin J. Fundamental issues in finite element analysis of localization of deformation. Engineering Computations 1993;10(2): 99e122.

27. Diederichs MS. Manuel Rocha medal recipient: rock fracture and collapse under low confinement conditions. Rock Mechanics and Rock Engineering 2003;36(5): 339e81.

28. Dyszlewicz J. Micropolar theory of elasticity. New York: Springer; 2004. Eberhardt E, Stead D, Coggan JS. Numerical analysis of initiation and progressive failure in natural rock slopes e the 1991

29. Randa rockslide. International Journal of Rock Mechanics and Mining Sciences 2004;41(1):69e87.

30. Eberhardt E, Stead D, Stimpson B, Read RS. Changes in acoustic event properties with progressive fracture damage. International Journal of Rock Mechanics and Mining Sciences 1997;34(3/4). 71.e1e71.e12.

31. Elmo D, Stead D. An integrated numerical modelling e discrete fracture network approach applied to the characterisation of rock mass strength of naturally fractured pillars. Rock Mechanics and Rock Engineering 2010;43(1):3e19.

32. Fakhimi A, Carvalho F, Ishida T, Labuz JF. Simulation of failure around a circular opening in rock. International Journal of Rock Mechanics and Mining Sciences 2002;39(4):507e15.

33. Fang Z, Harrison JP. Development of a local degradation approach to the modelling of brittle fracture in heterogeneous rocks. International Journal of Rock Mechanics and Mining Sciences 2002;39(4):443e57.

34. Feng XT, Pan PZ, Zhou H. Simulation of the rock microfracturing process under uniaxial compression using an elasto-plastic cellular automaton. International Journal of Rock Mechanics and Mining Sciences 2006;43(7):1091e108.

35. Ghaboussi J, Wilson EL, Isenberg J. Finite elements for rock joints and interfaces. Journal of the Soil Mechanics and Foundation Division, ASCE 1973;99(10): 849e62.

36. Goodman RE, Taylor RL, Brekke TL. A model for the mechanics of jointed rock. Journal of the Soil Mechanics and Foundation Division, ASCE 1968;94(4): 637e60.

37. Goodman RE. Introduction to rock mechanics. 2nd ed. New York: John Wiley & Sons Ltd.; 1989.

38. Hammah RE, Yacoub TE, Corkum B, Wibowo F, Curran JH. Analysis of blocky rock slopes with finite element shear strength reduction analysis. In: Eberhardt E, Stead D, Morrison T, editors. Proceedings of the 1st CanadaeUS Rock Mechanics Symposium. Vancouver, Canada; 2007. pp. 329e34.

39. Hammah RE, Yacoub T, Corkum B, Curran JH. The practical modelling of discontinuous rock masses with finite element analysis. In: Proceedings of the 42nd US Rock Mechanics Symposium and 2nd USeCanada Rock Mechanics Symposium. San Francisco, USA; 2008.

40. Harthong B, Scholtès L, Donzé F. Strength characterization of rock masses, using a coupled DEM-DFN model. Geophysical Journal International 2012;191(2): 467e80.

41. Hatzor YH, Benary R. The stability of a laminated voussoir beam: back analysis of a historic roof collapse using DDA. International Journal of Rock Mechanics and Mining Sciences 1998;35(2):165e81.

42. Hatzor YH, Wainshtein I, Mazor DB. Stability of shallow karstic caverns in blocky rock masses. International Journal of Rock Mechanics and Mining Sciences 2010;47(8):1289e303.

43. Hazzard JF, Young RP. Simulating acoustic emissions in bonded-particle models of rock. International Journal of Rock Mechanics and Mining Sciences 2000;37(5): 867e72.

44. Hazzard JF, Collins DS, Pettitt WS, Young RP. Simulation of unstable fault slip in granite using a bonded-particle model. Pure and Applied Geophysics 2002;159(1e3):221e45.

45. Hazzard JF, Young RP. Moment tensors and micromechanical models. Tectonophysics 2002;356(1e3):181e97.

46. Hazzard JF, Young RP. Dynamic modelling of induced seismicity. International Journal of Rock Mechanics and Mining Sciences 2004;41(8):1365e76.

47. Hillerborg A, Modéer M, Petersson PE. Analysis of crack formation and crack growth in concrete by means of fracture mechanics and finite elements. Cement and Concrete Research 1976;6(6):773e81.

48. Hoek E, Carranza-Torres CT, Corkum B. Hoek-Brown failure criterion e 2002 edition. In: Proceedings of the 5th North American Rock Mechanics Symposium. Toronto, Canada; 2002. pp. 267e73. 31.

49. Hoek E, Diederichs MS. Empirical estimates of rock mass modulus. International Journal of Rock Mechanics and Mining Sciences 2006;43(2):203e15.

50. Hoek E, Kaiser PK, Bawden WF. Support of underground excavations in hard rock. Taylor & Francis/A.A. Balkema; 1995.

51. Hoek E. Practical rock engineering. North Vancouver, Canada: Evert Hoek Consulting Engineer Inc.; 2006.

52. Hoek E. Strength of jointed rock masses. Geotechnique 1983;33(1):187e223.

53. Horii H, Nemat-Nasser S. Brittle failure in compression: splitting, faulting, and brittle-ductile transition. Philosophical Transactions of the Royal Society of London 1986;319(1549):337e74.

54. Itasca Consulting Group Inc.. FLAC e fast Lagrangian analysis of continua. Minneapolis, USA: Itasca Consulting Group Inc.; 2012a.

55. Itasca Consulting Group Inc.. PFC2D (particle flow code in 2 dimensions). Minneapolis, USA: Itasca Consulting Group Inc.; 2012b.

56. Itasca Consulting Group Inc.. UDEC (universal distinct element code). Minneapolis, USA: Itasca Consulting Group Inc.; 2013.

57. Jaeger JC, Cook NGW. Fundamentals of rock mechanics. 2nd ed. London: Chapman & Hall; 1976.

58. Jiang Y, Li B, Yamashita Y. Simulation of cracking near a large underground cavern in a discontinuous rock mass using the expanded distinct element method. International Journal of Rock Mechanics and Mining Sciences 2009;46(1):97e106.

59. Jing L, Hudson JA. Numerical methods in rock mechanics. International Journal of Rock Mechanics and Mining Sciences 2002;39(4):409e27.

60. Jing L, Stephansson O. Fundamentals of discrete element methods for rock engineering: theory and applications. Amsterdam/Oxford: Elsevier; 2007.

61. Jing L. A review of techniques, advances and outstanding issues in numerical modelling for rock mechanics and rock engineering. International Journal of Rock Mechanics and Mining Sciences 2003;40(3):283e353.

62. Kaiser PK, Kim BH. Rock mechanics challenges of underground constructions and mining. In: Proceedings of the Korean Rock Mechanics Symposium. Seoul, South Korea; 2008. pp. 1e6.

63. Karami A, Stead D. Asperity degradation and damage in the direct shear test: a hybrid FEM/DEM approach. Rock Mechanics and Rock Engineering 2008;41(2): 229e66.

64. Kazerani T, Yang ZY, Zhao J. A discrete element model for predicting shear strength and degradation of rock joint by using compressive and tensile test data. Rock Mechanics and Rock Engineering 2012;45(5):695e709.

65. Kazerani T, Zhao J. Micromechanical parameters in bonded particle method for modelling of brittle material failure. International Journal for Numerical and Analytical Methods in Geomechanics 2010;34(18):1877e95.

66. Kazerani T. Effect of micromechanical parameters of microstructure on compressive and tensile failure process of rock. International Journal of Rock Mechanics and Mining Sciences 2013;64:44e55.

67. Ke TC. Application of DDA to simulate fracture propagation in solid. In: Ohnishi Y, editor. Proceedings of the 2nd International Conference on Analysis of Discontinuous Deformation. Kyoto, Japan: Japanese Institute of Systems Research; 1997. pp. 155e85.

68. Kemeny J. Time-dependent drift degradation due to the progressive failure of rock bridges along discontinuities. International Journal of Rock Mechanics and Mining Sciences 2005;42(1):35e46.

69. Klerck PA, Sellers EJ, Owen DRJ. Discrete fracture in quasi-brittle materials under compressive and tensile stress states. Computer Methods in Applied Mechanics and Engineering 2004;193(27e29):3035e56.

70. Klerck PA. The finite element modelling of discrete fracture in quasi-brittle materials. PhD Thesis. Swansea, UK: University of Wales; 2000.

71. Koo CY, Chern JC. Modelling of progressive fracture in jointed rock by DDA method. In: Ohnishi Y, editor. Proceedings of the 2nd International Conference on Analysis of Discontinuous Deformation. Kyoto, Japan: Japanese Institute of Systems Research; 1997. pp. 186e201.

72. Kozicki J, Donzé FV. A new open-source software developed for numerical simulations using discrete modelling methods. Computer Methods in Applied Mechanics and Engineering 2008;197(49/50):4429e43.

73. Kozicki J, Donzé FV. YADE-OPEN DEM: an open-source software using a discrete element method to simulate granular material. Engineering Computations 2009;26(7):786e805.

74. Lan H, Martin C, Hu B. Effect of heterogeneity of brittle rock on micromechanical extensile behaviour during compression loading. Journal of Geophysical Research: Solid Earth 2010;115(B1):1e14.

75. Lemos JV. Recent development and future trends in distinct element methods e UDEC/3DEC and PFC codes. In: Zhao J, Ohnishi Y, Zhao G, Sasaki T, editors. Advances in Discontinuous Numerical Methods and Applications in Geomechanics and Geoengineering. London: Taylor & Francis Group; 2012.

76. Lin CT, Amadei B, Jung J, Dwyer J. Extensions of discontinuous deformation analysis for jointed rock masses. International Journal of Rock Mechanics and Mining Sciences & Geomechanics Abstracts 1996;33(7):671e94.

77. Lisjak A, Grasselli G, Vietor T. Continuum-discontinuum analysis of failure mechanisms around unsupported circular excavations in anisotropic clay shales. International Journal of Rock Mechanics and Mining Sciences 2014b;65:96e115.

78. Lisjak A, Liu Q, Zhao Q, Mahabadi OK, Grasselli G. Numerical simulation of acoustic emission in brittle rocks by two-dimensional finite-discrete element analysis. Geophysical Journal International 2013;195(3):423e43.

79. Lisjak A, Tatone BSA, Grasselli G, Vietor T. Numerical modelling of the anisotropic mechanical behaviour of Opalinus Clay at the laboratory-scale using FEM/DEM. Rock Mechanics and Rock Engineering 2014a;47(1):187e206.

80. Lisjak A. Investigating the influence of mechanical anisotropy on the fracturing behaviour of brittle clay shales with application to deep geological repositories. PhD Thesis. Toronto, Canada: University of Toronto; 2013.

81. Lockner DA, Byerlee JD, Kusenko V, Ponomarev A, Sidorin A. Quasi-static fault growth and shear fracture energy in granite. Nature 1991;350:39e42.

82. Lorig LJ, Cundall PA. Modeling of reinforced concrete using the distinct element method. In: Shah SP, Swartz SE, editors. Fracture of Concrete and Rock, SEMRILEM International Conference. Springer; 1989. pp. 276e87.

83. Ma GW, Wang XJ, Ren F. Numerical simulation of compressive failure of heterogeneous rock-like materials using SPH method. International Journal of Rock Mechanics and Mining Sciences 2011;48(3):353e63.

84. Mahabadi OK, Cottrell BE, Grasselli G. An example of realistic modelling of rock dynamics problems: FEM/DEM simulation of dynamic Brazilian test on Barre granite. Rock Mechanics and Rock Engineering 2010a;43(6):707e16.

85. Mahabadi OK, Lisjak A, Grasselli G, Lukas T, Munjiza A. Numerical modelling of a triaxial test of homogeneous rocks using the combined finite-discrete element method. In: Zhao J, Labiouse V, Dudt J, Mathier J, editors. Proceedings of the European Rock Mechanics Symposium (EUROCK) 2010. A.A. Balkema; 2010b.

86. Mahabadi OK, Grasselli G, Munjiza A. Numerical modelling of a Brazilian disc test of layered rocks using the combined finite-discrete element method. In: Diederichs M, Grasselli G, editors. Proceedings of

the 3rd CanadaeUS (CANUS) Rock Mechanics Symposium (RockEng09). Toronto, Canada; 2009.

87. Mahabadi OK, Lisjak A, Munjiza A, Grasselli G. Y-Geo: new combined finite-discrete element numerical code for geomechanical applications. International Journal of Geomechanics 2012a;12:676e88.

88. Mahabadi OK, Randall NX, Zong Z, Grasselli G. A novel approach for micro-scale characterization and modelling of geomaterials incorporating actual material heterogeneity. Geophysical Research Letters 2012b;39(1). http://dx.doi.org/ 10.1029/2011GL050411.

89. Mahabadi OK. Investigating the influence of micro-scale heterogeneity and microstructure on the failure and mechanical behaviour of geomaterials. PhD Thesis. Toronto, Canada: University of Toronto; 2012.

90. Manouchehrian A, Marji MF. Numerical analysis of confinement effect on crack propagation mechanism from a flaw in a pre-cracked rock under compression. Acta Mechanica Sinica 2012;28(5):1389e97.

91. Martin CD, Read RS, Martino JB. Observations of brittle failure around a circular test tunnel. International Journal of Rock Mechanics and Mining Sciences 1997; 34(7):1065e73.

92. Martin CD. Seventeenth Canadian Geotechnical Colloquium: the effect of cohesion loss and stress path on brittle rock strength. Canadian Geotechnical Journal 1997;34(5):698e725.

93. Mas Ivars D, Pierce ME, Darcel C, Reyes-Montes J, Potyondy DO, Young RP, Cundall PA. The synthetic rock mass approach for jointed rock mass modelling. International Journal of Rock Mechanics and Mining Sciences 2011;48(2): 219e44.

94. Mas Ivars D, Potyondy DO, Pierce M, Cundall PA. The smooth-joint contact model. In: Proceedings of the 8th World Congress on Computational Mechanics e 5th European Congress on Computation Mechanics and Applied Science and Engineering. Venice, Italy; 2008.

95. Moon T, Nakagawa M, Berger J. Measurement of fracture toughness using the distinct element method. International Journal of Rock Mechanics and Mining Sciences 2007;44(3):449e56.

96. Mühlhaus HB, Aifantis EC. A variational principle for gradient plasticity. International Journal of Solids and Structures 1991;28(7):845e57.

97. Mühlhaus HB, Vardoulakis I. The thickness of shear bands in granular materials. Geotechnique 1987;37(3):271e83.

98. Mühlhaus HB. Continuum models for layered and blocky rock. In: Hudson JA, editor. Comprehensive rock engineering. Oxford: Pergamon Press; 1993. pp. 209e30.

99. Munjiza A, Andrews KRF, White JK. Combined single and smeared crack model in combined finite-discrete element analysis. International Journal for Numerical Methods in Engineering 1999;44(1):41e57.

100. Munjiza A, Andrews KRF. NBS contact detection algorithm for bodies of similar size. International Journal for Numerical Methods in Engineering 1998;43(1):131e49.

101. Munjiza A, Andrews KRF. Penalty function method for combined finiteediscrete element systems comprising large number of separate bodies. International Journal for Numerical Methods in Engineering 2000;49(11):1377e96.

102. Munjiza A, John NWM. Mesh size sensitivity of the combined FEM/DEM fracture and fragmentation algorithms. Engineering Fracture Mechanics 2002;69(2): 281e95.

103. Munjiza A, Owen DRJ, Bicanic N. A combined finite-discrete element method in transient dynamics of fracturing solids. Engineering Computations 1995;12(2): 145e74.

104. Munjiza A. The combined finite-discrete element method. Chichester, UK: John Wiley & Sons Ltd.; 2004.

105. Owen DRJ, Feng YT. Parallelised finite/discrete element simulation of multifracturing solids and discrete systems. Engineering Computations 2001;18(3/ 4):557e76.

106. Pan XD, Reed MB. A coupled distinct element-finite element method for large deformation analysis of rock masses. International Journal of Rock Mechanics and Mining Sciences & Geomechanics Abstracts 1991;28(1):93e9.

107. Pande GN, Sharma KG. On joint/interface elements and associated problems of numerical ill-conditioning. International Journal for Numerical and Analytical Methods in Geomechanics 1979;3(3):293e300.

108. Paterson MS, Wong T. Experimental rock deformation e the brittle field. New York: Springer; 2004.

109. Pine RJ, Coggan JS, Flynn ZN, Elmo D. The development of a new numerical modelling approach for naturally fractured rock masses. Rock Mechanics and Rock Engineering 2006;39(5):395e419.

110. Pine RJ, Owen DRJ, Coggan JS, Rance JM. A new discrete fracture modelling approach for rock masses. Geotechnique 2007;57(9):757e66.

111. Potyondy DO, Cundall PA. Modeling notch-formation mechanisms in the URL mineby test tunnel using bonded. International Journal of Rock Mechanics and Mining Sciences 1998;35(4/5):510e1.

112. Potyondy DO, Cundall P. Bonded-particle simulations of the in-situ failure test at Olkiluoto. International Progress Report 01-13. Stockholm, Sweden: SKB; 2000.

113. Potyondy DO, Cundall PA, Lee C. Modeling of rock using bonded assemblies of circular particles. In: Aubertin M, editor. Proceedings of the 2nd North American Rock Mechanics Symposium e NARMS'96. Brookfield, USA: A.A. Balkema; 1996. pp. 1934e44.

114. Potyondy DO, Cundall PA. A bonded-particle model for rock. International Journal of Rock Mechanics and Mining Sciences 2004;41(8):1329e64.

115. Potyondy DO. A flat-jointed bonded-particle material for hard rock. In: Proceedings of the 46th US Rock Mechanics/Geomechanics Symposium. Chicago, USA: American Rock Mechanics Association; 2012.

116. Rasouli V, Harrison J. Assessment of rock fracture surface roughness using Riemannian statistics of linear profiles. International Journal of Rock Mechanics and Mining Sciences 2010;47(6):940e8.

117. Riahi A, Hammah ER, Curran JH. Limits of applicability of the finite element explicit joint model in the analysis of jointed rock problems. In: Proceedings of the 44th U.S.

118. Rock Mechanics Symposium and 5th USeCanada Rock Mechanics Symposium. Salt Lake City, USA; 2010.

119. Rockfield Software Ltd.. ELFEN 2D/3D numerical modelling package. Swansea, UK: Rockfield Software Ltd.; 2004.

120. Rougier E, Knight EE, Sussman AJ, Swift RP, Bradley CR. The combined finite-discrete element method applied to the study of rock fracturing behaviour in 3D. In: Proceedings of the 45th US Rock Mechanics/Geomechanics Symposium. San Francisco, USA: American Rock Mechanics Association; 2011.

121. Sagong M, Park D, Yoo J, Lee JS. Experimental and numerical analyses of an opening in a jointed rock mass under biaxial compression. International Journal of Rock Mechanics and Mining Sciences 2011;48(7):1055e67.

122. Scholtès L, Donzé F, Khanal M. Scale effects on strength of geomaterials, case study: coal. Journal of the Mechanics and Physics of Solids 2011;59(5):1131e46.

123. Scholtès L, Donzé F. Modelling progressive failure in fractured rock masses using a 3D discrete element method. International Journal of Rock Mechanics and Mining Sciences 2012;52:18e30.

124. Scholtès L, Donzé F. A DEM model for soft and hard rocks: role of grain interlocking on strength. Journal of the Mechanics and Physics of Solids 2013;61(2):352e69.

125. Sellers EJ, Klerck PA. Modelling of the effect of discontinuities on the extent of the fracture zone surrounding deep tunnels. Tunnelling and Underground Space Technology 2000;15(4):463e9.

126. Shi G, Goodman R. Discontinuous deformation analysis e a new method for computing stress, strain and sliding of block systems. In: Proceedings of the 29th US Symposium on Rock Mechanics. Rotterdam: A.A. Balkema; 1988.

127. Smilauer V, Catalano E, Chareyre B, Dorofenko S, Duriez J, Gladky A, Kozicki J, Modenese C, Scholtès L, Sibille L, Stránský J, Thoeni K. Yade documentation. URL: http://yade-dem.org/doc/; 2010.

128. Stefanizzi S, Barla G, Kaiser PK, Grasselli G. Numerical modeling of rock mechanics tests in layered media using a finite/discrete element approach. In: Proceedings of the 12th International Conference of International Association for Computer Methods and Advances in Geomechanics (IACMAG). Goa, India; 2008.

129. Stefanizzi S. Numerical modelling of strain-driven fractures around tunnels in layered rock masses. PhD Thesis. Turin, Italy: Politecnico di Torino; 2007.

130. Tang C. Numerical simulation of progressive rock failure and associated seismicity. International Journal of Rock Mechanics and Mining Sciences 1997;34(2): 249e61.

131. Tang CA, Lü HY. The DDD method based on the combination of RFPA and DDA. In: Chen G, Ohnishi Y, Zheng L, Sasaki T, editors. Frontiers of Discontinuous Numerical Methods and Practical Simulations in Engineering and Disaster Prevention. London: Taylor & Francis Group; 2013. pp. 105e12.

132. Tannant DD, Wang C. Thin tunnel liners modelled with particle flow code. Engineering Computations 2004;21(2e4):318e42.

133. Vyazmensky A, Elmo D, Stead D. Role of rock mass fabric and faulting in the development of block caving induced surface subsidence. Rock Mechanics and Rock Engineering 2010a;43(5):533e56.

134. Vyazmensky A, Stead D, Elmo D, Moss A. Numerical analysis of block cavinginduced instability in large open pit slopes: a finite element/discrete element approach. Rock Mechanics and Rock Engineering 2010b;43(1):21e39.

135. Wanne TS, Young RP. Bonded-particle modelling of thermally fractured granite. International Journal of Rock Mechanics and Mining Sciences 2008;45(5): 789e99.

136. Wanne TS. Bonded-particle simulation of tunnel sealing experiment. In: Diederichs M, Grasselli G, editors. ROCKENG09: Proceedings of the 3rd CANUS Rock Mechanics Symposium. Toronto, Canada; 2009.

137. Wilson EL. Finite elements for foundations, joints and fluids. In: Gudehus G, editor. Finite Elements in Geomechanics. Chichester, UK: John Wiley & Sons Ltd.; 1977. Yan M. Numerical modelling of brittle fracture and step-path failure: from laboratory to rock slope scale. PhD Thesis. Burnaby, Canada: Simon Fraser University; 2008.

138. Yuan SC, Harrison JP. A review of the state of the art in modelling progressive mechanical breakdown and associated fluid flow in intact heterogeneous rocks. International Journal of Rock Mechanics and Mining Sciences 2006;43(7): 1001e22.

139. Zhang XP, Wong LNY. Cracking processes in rock-like material containing a single flaw under uniaxial compression: a numerical study based on bonded-particle model approach. Rock Mechanics and Rock Engineering 2012;45(5):711e37.

140. Zhang XP, Wong LNY. Crack initiation, propagation and coalescence in rock-like material containing two flaws: a numerical study based on bonded-particle model approach. Rock Mechanics and Rock Engineering 2013;46(5):1001e21.

141. Zhao Z. Gouge particle evolution in a rock fracture undergoing shear: a microscopic DEM study. Rock Mechanics and Rock Engineering 2013;46(6):1461e79.

CITATION

A. Lisjak, G. Grasselli, A review of discrete modeling techniques for fracturing processes in discontinuous rock masses, Journal of Rock Mechanics and Geotechnical Engineering, Volume 6, Issue 4, August 2014, Pages 301-314, ISSN 1674-7755, http://dx.doi.org/10.1016/j.jrmge.2013.12.007.

CHAPTER 3

3D Random Voronoi Grain-Based Models for Simulation of Brittle Rock Damage and Fabric-Guided Micro-Fracturing

E. Ghazvinian[a], M.S. Diederichs[a], R. Quey[b]

[a] GeoEngineering Centre, Queen's University, Kingston, ON, Canada

[b] Ecole Nationale Supérieure des Mines de Saint-Etienne, CNRS UMR 5307, France

ABSTRACT

A grain-based distinct element model featuring three-dimensional (3D) Voronoi tessellations (random poly-crystals) is proposed for simulation of crack damage development in brittle rocks. The grain boundaries in poly-crystal structure produced by Voronoi tessellations can represent flaws in intact rock and allow for numerical replication of crack damage progression through initiation and propagation of micro-fractures along grain boundaries. The Voronoi modelling scheme has been used widely in the past for brittle fracture simulation of rock materials. However the difficulty of generating 3D Voronoi models has limited its application to two-dimensional (2D) codes. The proposed approach is implemented in Neper, an open-source engine for generation of 3D Voronoi grains, to generate block geometry files that can be read directly into 3DEC. A series of Unconfined Compressive Strength (UCS) tests are simulated in 3DEC to verify the proposed methodology for 3D simulation of brittle fractures and to investigate the relationship between each micro-parameter and the model's macro-response. The possibility of numerical replication of the classical U-shape strength curve for anisotropic rocks is also investigated in numerical UCS tests by using complex-shaped (elongated) grains that are cemented to one another along their adjoining sides. A micro-parameter calibration procedure is established for 3D Voronoi models for accurate replication of the mechanical behaviour of isotropic and anisotropic (containing a fabric) rocks.

INTRODUCTION

The simulation of micro-fracturing and crack damage progression for brittle rocks can be performed implicitly in continuum or explicitly with the aid of discontinuum numerical approaches. In implicit simulation of micro-crack formation, the weakness caused by the formation of cracks is smeared within the material by means of constitutive relationships. The Damage Initiation-Spalling Limit (DISL) approach is an example of implicit brittle fracture simulation with continuum codes (Diederichs, 2007). The explicit simulation of micro-fracturing in rock-like material can be accomplished by direct representation of cracks in models formulated based on the Discrete Element Method (DEM) (Cundall and Hart, 1985 and Cundall, 1988) or by using the hybrid Finite-Discrete Element Method (FDEM) codes (Mahabadi et al., 2012).

In DEM, the rock-like material can be simulated as a dense assembly of rigid or deformable particles that interact at their contact points. Also, the discrete bodies can detach and new contacts can automatically be detected (Itasca, 2013a) and therefore fractures can be simulated at the grain (or block) boundary opening. In the Bonded-Particle Model (BPM), which falls within the DEM formulation, the particles are represented by rigid disks in

2D and rigid spheres in 3D (Potyondy and Cundall, 2004). The breakage of bonds between the spheres in the BPM is comparable to fracturing in rock. The BPM which is implemented in Particle Flow Code (PFC) (Itasca, 2008) has been used widely for crack damage simulation in rock. The concerns and possible solutions for the accuracy of PFC simulations have been documented in the past by many researchers, including Diederichs, 2003, Potyondy and Cundall, 2004, Cho et al., 2007 and Yoon et al., 2008, and Ghazvinian (2010).

The random polygonal blocks (grains) in DEM are an alternative geometry to the disks and spheres employed in BPM approach for simulation of fracturing. Voronoi tessellation is one of the available techniques for generating the random polygonal grains within a domain. In this technique, a region is populated with random seed points. Lines or planar surfaces are generated so that the bounded region surrounding each seed point includes all the space that is closer to that seed point than any other. The grain boundaries in the poly-crystal structure produced by Voronoi tessellation can be used to represent flaws in intact rock and therefore allow for simulation of crack damage development through initiation and propagation of fractures along grain boundaries.

In the Voronoi modelling approach for numerical simulation of fracturing in intact rock, the grains (blocks) can be rigid, deformable or inelastic. The behaviour of contacts between the grains is commonly governed by the available constitutive formulations for rock mass joints and discontinuities. Some of the concerns associated with the PFC results were demonstrated to be no longer applicable to the Voronoi method (Kazerani and Zhao, 2010 and Lan et al., 2010). The Voronoi scheme has been used widely in the past for brittle fracture modelling of rock materials, however the complexity of generating 3D Voronoi domains has limited its application to 2D models. In this paper, a new methodology for generation of 3D Voronoi Grain-Based Models (GBMs) is proposed which would provide a means for simulation of rock brittle fracturing in 3DEC (Itasca, 2013a).

Damage process in laminated intact rocks or rocks that have a preferred fabric orientation, which commonly represent an anisotropic mechanical behaviour, can also be simulated by means of an equivalent continuum such as the Ubiquitous Joint Model in FLAC (Itasca, 2012), or with DEM and hybrid FDEM discontinuum approaches. The discontinuum methods either apply implicit anisotropic contact and material constitutive behaviour or introduce an anisotropic defect or pre-damage to the model (e.g. Diederichs et al., 2004, You et al., 2011, Kazerani, 2013 and Lisjak et al., 2014a). By using the proposed approach in this paper for generation of 3D Voronoi GBMs, an analogue for intact rocks with inherent anisotropy is suggested by scaling the models and therefore achieving elongated grains. The fabric-guided micro-fracturing in the anisotropic models is verified by laboratory test results of the Cobourg Limestone and the calibration procedure for the laminated GBMs is established from sensitivity analyses of the contact micro-parameters.

NUMERICAL APPROACH

The extensile "crack opening forces" at the micro-scale can be generated through different mechanisms. These mechanisms amplify the localized shear strains and subsequently facilitate the nucleation of inter-granular and intra-granular cracks at the Crack Initiation (CI) threshold (Diederichs, 2003). Some of the mechanisms of extension crack generation under deviatoric stress conditions are illustrated in Fig. 1.

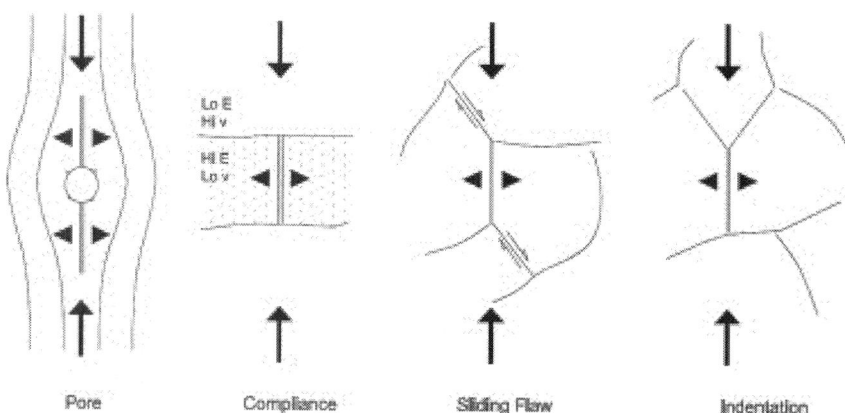

Figure 1. Schematic illustration of crack opening forces generated through different mechanisms at the micro-scale under deviatoric stress condition. High (Hi) and low (Lo) E and v are the Young's modulus and Poisson's ratio, respectively (after Diederichs (2000)).

To simulate the extensile crack opening forces at the micro-scale in DEM models, the rigid or deformable particles are approximated by simple geometries. The three commonly used geometries are disks (spheres in 3D) (Potyondy and Cundall, 2004 and Scholtès and Donzé, 2013), Voronoi grains (Damjanac et al., 2007), and triangles (trigons in 3D) (Gao and Stead, 2014). The geometry of the particles plays a key role in the generation of the crack extension and shearing forces acting on the sliding fractures. A schematic of an identical micro-mechanism simulated with Voronoi grains and circular disks is illustrated in Fig. 2. The applied compressive forces in the Voronoi model can be resolved into extensile and shearing forces acting on the grain boundaries (sliding flaw in this case). In the circular disk model, the applied forces will translate into tensile stresses at the contact between the disks and rotational moments acting upon the disks. The geometrically imposed restriction on the generation of shear stresses at the contacts between the disks in the BPM has a minor effect on the CI threshold since most of the micro-cracks in brittle rocks form in tension (Diederichs, 2007). However, with increasing deviatoric stress and on reaching the Critical Damage (CD) threshold, micro-cracks start to interact and coalesce, mostly through shearing. Therefore, the BPM with circular disks has limited application for simulation of crack interaction (Diederichs, 2007).

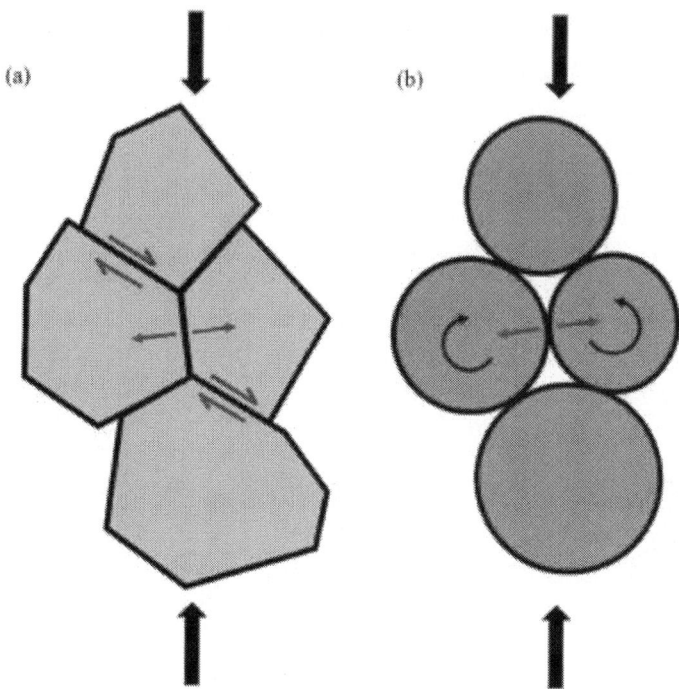

Figure 2. Micro-mechanisms for compression-induced tensile stress at the contacts between the particles in DEM. (a) The compressive force can be resolved into tensile and shear stresses acting on the boundaries between the grains. (b) The compressive forces are resolved into tensile stress at the contacts and moments trying to rotate the disks.

The simplicity of generating triangular grains (trigons) in DEM for simulation of fracturing in rock is assisting the increasing rate of using this approach. The micro-properties of the trigons models can be calibrated to simulate the exact macro-properties of rock tested in the laboratory. At the micro-scale, in contrary to the BPM models, most of the failure occurs in shear (Gao and Stead, 2014). In Fig. 3a, the potential sliding path in a trigon model is shown (red lines). In Fig. 3b, a Voronoi tessellated model with similar edge length to the trigon model is also illustrated. While the trigon model includes smooth pathways that encourage shear sliding, the grain boundaries of the Voronoi model provide asperities and rough failure paths which lead to extensile opening of some grain boundaries and simultaneous shear sliding along other grain edges.

This figure shows that the existence of these smooth pathways for cracking encourages shear sliding.

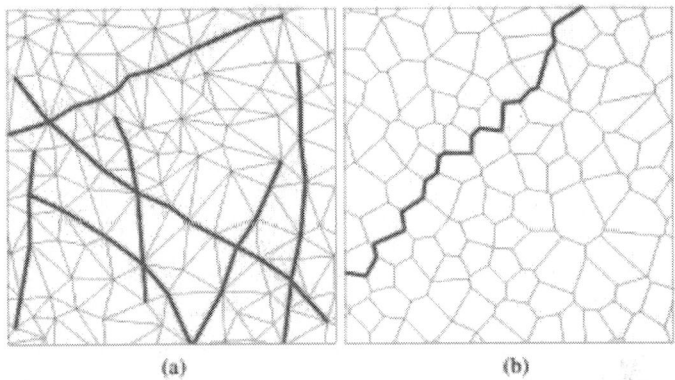

(a)   (b)

Figure 3. Smooth versus rough sliding paths for shearing in the (a) trigon and (b) Voronoi models, respectively. The potential sliding paths are shown as red lines.

MICROSTRUCTURE MODELLING FOR DEM

3D Voronoi tessellations

A 3D Voronoi tessellation is a partition of a domain of 3D space, $D \in R^3$, into a collection of cells. Given a number of seed points in D, $\{Si(xi)\}$ for $i = \{1, ..., N\}$, every seed is assigned a Voronoi cell, Ci, as follows:

$$C_i = \{P(x) \in D \mid d(P, S_i) \leq d(P, S_j) \quad \forall j \neq i\} \tag{1}$$

where $d(\bullet, \bullet)$ is the Euclidean distance. The seed locations are randomly chosen from a uniform distribution in the domain. A Voronoi cell can be seen as the region of influence of a seed, namely the region of space closer to the seed than to any other seed. Voronoi cells are convex polyhedra intersecting along flat faces, straight edges and vertices, for 2, 3 and 4 cells, respectively. An example of a Voronoi tessellation containing 2000 cells is provided in Fig. 4a. Different algorithms have been proposed for constructing Voronoi tessellations. In the present work, a cell-by-cell construction algorithm is used. A cell (Ci, of seed Si) is first set as the whole domain. Ci is then modified by an iterative process, where other seeds (Sj) are considered by increasing distance from Si. At each iteration (seed Sj), Ci is reduced to the intersection of the previously computed cell and the half-space closer to Si than to Sj. The iterative process can be stopped when the distance between seed Si and seed Sj becomes high enough for the half-space closer to Si than to Sj to necessarily include the whole cell. A simple, isotropic criterion is

$$\left. \begin{array}{l} d(S_j, S_i) > 2d_{\max} \\ d_{\max} = \max_{P \in C_i} d(P, S_i) = \max_{P \in \{V_i\}} d(P, S_i) \end{array} \right\} \tag{2}$$

where $\{Vi\}$ is the set of vertices of cell Ci. The resulting collection of cells fully defines the Voronoi tessellation. However, cell interfaces are defined several times: a cell face, edge or vertex is defined in 2, 3 or 4 cells, respectively. Therefore, for a proper description of a tessellation, duplicate features are merged. In general, this is also needed for further processing, for example regularization (see Section 3.2) and meshing (see Quey et al. (2011)).

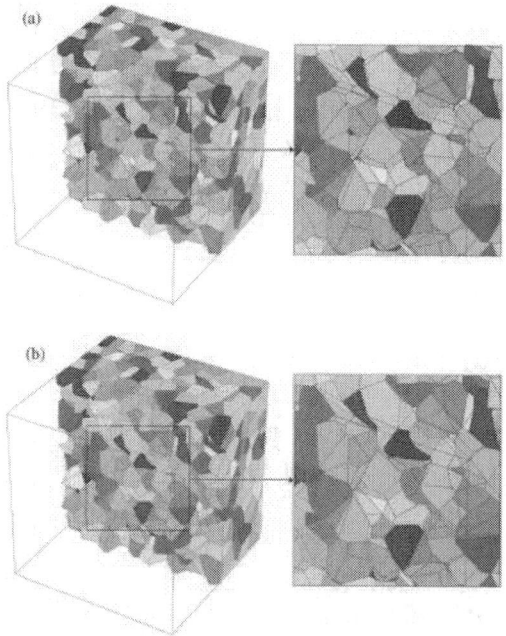

Figure 4. Effect of regularization illustrated on a 2000-cell tessellation. (a) Standard Voronoi tessellation and (b) regularized Voronoi tessellation. Only half of the cells are shown and the small edges are plotted in red. 8000 out of 25,000 edges are removed by regularization.

The interest of Voronoi tessellations for microstructure modelling is twofold. First, similarly to regular 3D tessellations, based on cubes, rhombic dodecahedra or truncated octahedra, Voronoi tessellations can be described in a vectorial format (contrary to a raster format as provided by tomography techniques or different modelling approaches). Such a compact description is particularly well-suited for DEM and can be meshed for other methods. Second, Voronoi tessellations show a wide variety of cell shapes and volumes, and spatial orientations of the contact surfaces between cells. This guarantees that the microstructure behaviour is not biased by particular cell arrangements or interface spatial orientations. This is why 3D Voronoi tessellations have been often used for finite element method (FEM) or boundary element method (BEM) simulations, e.g.Barbe et al., 2009, Barbe and Quey, 2011, Quey et al., 2012 and Benedetti and Aliabadi, 2013.

Regularization

As can be seen in Fig. 4a, Voronoi tessellations contain a high proportion of edges and faces of dimension significantly lower than the average cell size. Such features are negligible from a cell morphology point of view and therefore should not play a significant role in the results of the DEM simulations either. They also are highly detrimental to computation time in DEM simulations.

These drawbacks can be circumvented by "regularization", a technique consisting of removing the "small edges" of a tessellation, of length typically lower than a tenth of the cell size (Quey et al., 2011). Regularization removes small faces indirectly, because the latter are composed of one or several small edges. During regularization, the small edges are removed by increasing length. An edge (and its two vertices) is collapsed to a single vertex on the condition that the resulting geometrical distortion in the vicinity of the edge is sufficiently low (typically, the neighbouring faces must remain planar within a 200 angular tolerance). Constraints are imposed on the vertices, edges and faces at the boundary of the domain so that the domain shape is not changed. As illustrated inFig. 4b, regularization removes the small edges and faces while retaining the global cell shapes. In most cases, face distortions remain lower than 1°–2° (Quey et al., 2011). In 3DEC simulations, such distorted faces can be treated as standard, planar faces.

Two-scale Voronoi tessellations

A "two-scale Voronoi tessellation" is a tessellation for which every cell of a first, "primary" Voronoi tessellation, is partitioned by another, "secondary" Voronoi tessellation. Fig. 5provides two possible applications of two-scale Voronoi tessellations. First, such tessellations enable to model microstructures that clearly exhibit two scales such as sedimentary rocks, consisting of 45°-tilted bands decomposed into individual blocks (Fig. 5a). The primary Voronoi tessellation was created from seeds regularly spaced along a line (hence the band structure) and the secondary Voronoi tessellations were obtained using random seed locations. Second, two-scale tessellations enable inclusion of potential intra-grain crack paths into a regular, single-scale material (Fig. 5b). Intra-grain interfaces can be assigned micro-properties different from grain-to-grain interfaces. The algorithm for constructing a two-scale Voronoi tessellation consists of sequentially constructing standard, single-scale Voronoi tessellations. Secondary Voronoi tessellations can be constructed independently from each other. The Voronoi tessellation algorithm described in Section 4.1 is used, using the primary cell (instead of the whole domain) as a starting point for the secondary cells. This results in a collection of secondary tessellations. Similarly to single-scale Voronoi tessellations, duplicated, or overlapping features are found at the intersections between secondary tessellations. These features must be merged to get a proper description of the tessellation. The procedure involves merging, creation and decomposition of vertices, edges and faces located at the intersection between secondary tessellations. Details will be provided separately. Similar to single-scale tessellations, the resulting two-scale tessellations contain small edges and faces, which can be removed by regularization.

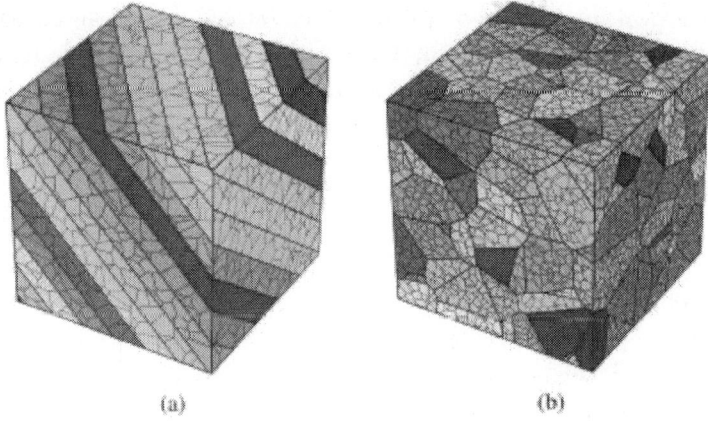

(a)   (b)

Figure 5. Two-scale Voronoi tessellations applied to modelling (a) a two-scale material (e.g. sedimentary rocks) and (b) intra-grain cracking paths.

The Neper software

All algorithms described previously have been implemented in a free (open-source) software: Neper (Quey et al., 2011 and Quey, 2014). The software offers support for both single-scale and two-scale Voronoi tessellations, in 2D and 3D, however the focus of this paper is on single-scale 3D Voronoi tessellated models. Thanks to efficient and robust algorithms, very large tessellations can be easily created (typically 105 cells). Any convex domain shape can be used; cubes (or any parallelepiped) and cylinders are directly supported. The tessellations can also be stretched to get elongated cells. Rotation and cropping are also straightforward operations. Finally, tessellations can be written at various formats, including 3DEC for DEM simulations, and meshed for FEM simulations.

GRAIN-BASED MODEL FOR 3DEC

The Voronoi blocks can be used for simulating fractures at different scales, from micro-cracks within a laboratory specimen (Lan et al., 2010) to modelling spalling and slabbing in the walls of underground excavations (Itasca, 2011). The focus of this paper is on GBMs. With the currently available computational power, GBMs are commonly restricted to the laboratory specimen scale. To investigate the application of the 3D Voronoi tessellation approach for simulation of brittle micro-fracturing in rock, a series of Uniaxial Compressive Strength (UCS) tests are simulated in 3DEC from the tessellations generated in Neper.

Specimen generation

Cylindrical specimens were decided to be used for the simulation of numerical UCS tests. Any 3D convex domain geometry can be filled with 3D Voronoi diagrams with Neper. To avoid grains whose geometry is affected by the domain boundary in a cylindrical specimen, a cubical domain was generated which was then carved into a cylindrical specimen in 3DEC (Fig. 6). For generation of the specimens, a cylinder with the height of 140 mm was cut from the centre of the tessellated block and then two 10 mm slices were cut from each end of the specimen. The grains in these 10 mm thick slices were joined to form the platens. The final specimen had a diameter of 55 mm and a length of 120 mm. The cubical domains in Neper were generated with 40,000 grains while the final cylindrical specimens contained 5300 grains on average.

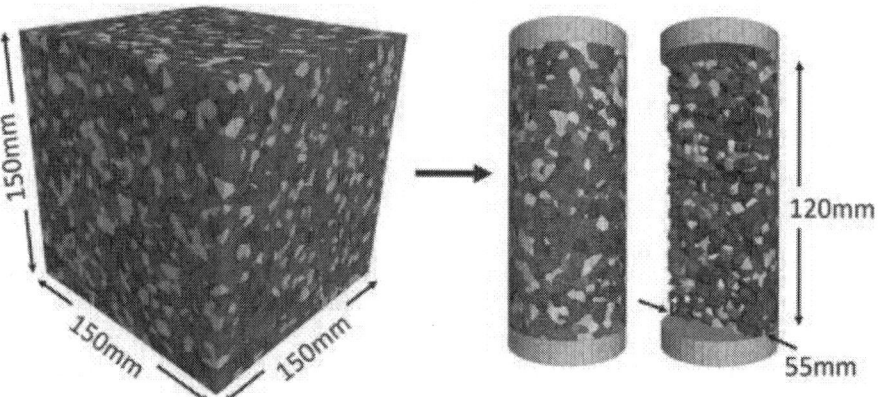

Figure 6. Generation of a Voronoi tessellated cube in Neper and cutting the cube into a cylindrical specimen in 3DEC.

Numerical test set-up

Loading of the UCS models was performed by applying a constant velocity to the top and bottom platens until a moderate state of monotonically increasing axial stress was reached in the specimens. After this stage, a servo-controlled system took over the loading control. This servo mechanism kept the average unbalanced mechanical forces in the model within a specified range by adjusting the applied velocity to the platens. The unbalanced force servo system is particularly important at the failure stage to keep the loading condition within the static range. The loading velocity at the beginning of the test and the specified range for unbalanced forces for the servo-mechanically controlled loading stage was kept small enough to ensure the uniform distribution of stress within the specimen.

The axial strain for the specimens was measured by averaging the displacements between five pairs of grid points, located 10 mm away from each end of the specimens in the vicinity of the models' axial axis, in principle similar to laboratory extensometers. The locations of these grid points are illustrated in Fig. 7. The specimens' lateral strain was calculated from the average relative displacement of three pairs of grid points in the middle of the specimens as shown in Fig. 7, in the x and y directions.

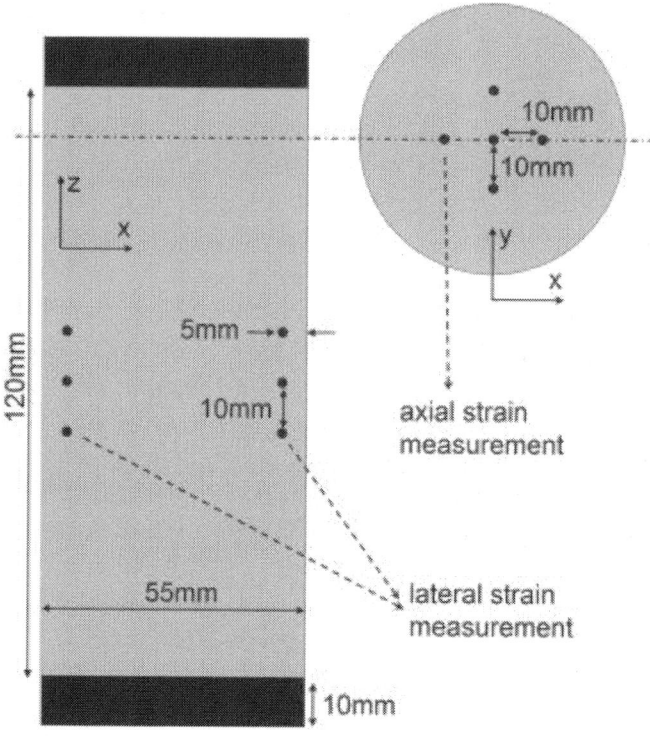

Figure 7. Location of grid points used to measure axial and lateral strains.

The axial stress in the specimens was measured by averaging the axial stress in the zones of blocks whose centroid falls within a cube in the centre of the cylindrical specimen. A trial test was run to determine the suitable size for the cube which was representative of the axial stress within the specimen. For this purpose, the axial stress within four boxes with dimensions shown in Fig. 8a was measured during a UCS test. The results are shown in Fig. 8b. It can be seen that the difference between the measured stresses tends to disappear with increasing box size, consistent with the emergence of a representative volume element (Kanit et al., 2003). Although the stress measured in the two largest cubes appear to be very similar, it was decided to use the biggest box size with dimensions of 35 mm × 35 mm × 40 mm to measure axial stress within the specimens. An example for the zones used for measurement of axial stress in a specimen is shown in Fig. 8c.

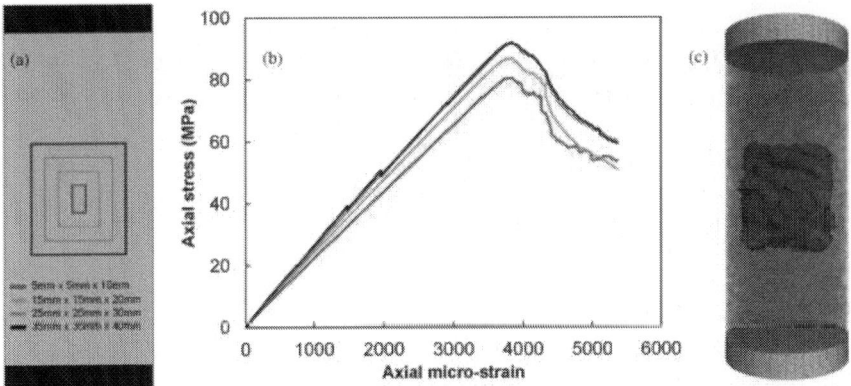

Figure 8. Measurement of axial stress for the specimens. (a) The four box sizes investigated for axial stress measurement. (b) Comparison between the measured axial stress within the four boxes of different dimensions. (c) Axial stress for the specimen is measured as the mean axial stress recorded in the red zones.

Estimation of crack damage thresholds

Estimation of the crack damage thresholds for the models was performed according to the Acoustic Emission (AE) method described in Ghazvinian et al. (2012). The comparable phenomenon to AE activity in laboratory specimens can be easily correlated to the bond breakage in PFC models. However, the same is not true for GBMs in 3DEC. Contact deletion, sub-contact deletion, sub-contact failure, area of the failed sub-contact, etc., can all be associated with the AE activity. In this paper, the number of failed sub-contacts was used, as shown in Fig. 9, to identify the damage thresholds. Nevertheless further investigation is required on this subject.

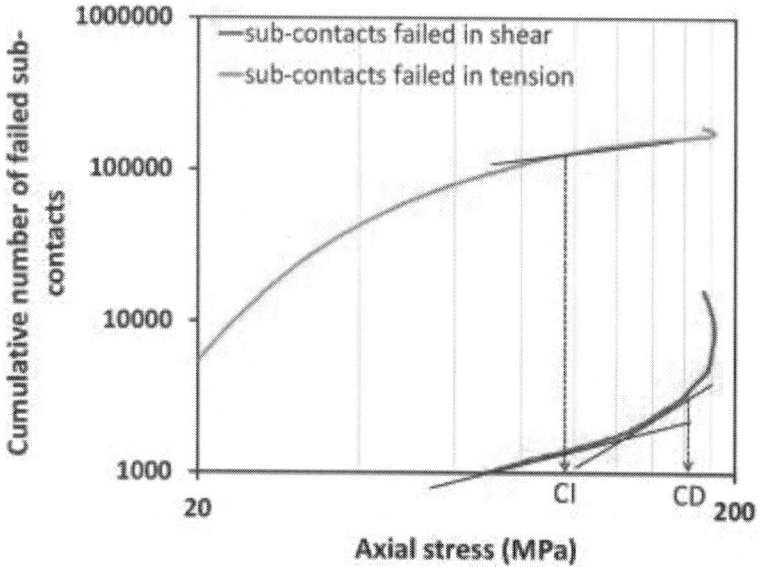

Figure 9. Estimation of crack damage thresholds from the number of failed sub-contacts in 3DEC.

GBM calibration

The behaviours of 3D Voronoi tessellated models are controlled by two main components: the grains (3DEC blocks) and the contacts between grains. The grains in a 3DEC model can be simulated as rigid bodies, either elastic or plastic, where continuum intra-grain failure is allowed to occur. Common practice with Voronoi GBMs is to have rigid (Kazerani and Zhao, 2010) or elastic grains (Itasca, 2011) for simplicity and to allow for failure to occur only at contacts between the grains. In this case, the parameters requiring calibration, when a simple Coulomb slip joint model is used for the grain contacts in 3DEC, are as follows:

- contact normal stiffness (k_n)
- contact normal to shear stiffness ratio (k_n/k_s)
- contact peak cohesion (c_p)
- contact residual cohesion (c_r)
- contact peak tensile strength (T_p)
- contact residual tensile strength (T_r)
- contact peak friction angle (ϕ_p)

- contact residual friction angle (ϕ_r)
- grain (block) Young's modulus (E)
- grain (block) Poisson's ratio (v)

For simplicity in the calibration process, the residual tensile strength (T_r) and cohesion (c_r) are commonly assumed to be zero (Kazerani and Zhao, 2010, Itasca, 2011 and Gao and Stead, 2014) for Voronoi GBMs.

Martin (1997) showed that the mobilization of the frictional and cohesional components of shearing in brittle rocks is not a simultaneous process. The friction gradually mobilizes as a function of damage to the brittle rock, while the cohesion is degrading.Hajiabdolmajid et al. (2002) successfully implemented the Cohesional Weakening-Frictional Strengthening (CWFS) behaviour in their continuum modelling to simulate notch formation around the tunnels of deep underground openings in massive brittle rocks. To adopt the CWFS concept for the contact behaviour between the grains, the peak cohesion (c_p) and residual friction angle (ϕ_r) were defined for the contacts while the residual cohesion (c_r) and peak friction angle (ϕ_p) were set to zero. Therefore for simplicity from now on in this paper, the peak cohesion, peak tensile strength and residual friction angle will be referred to as c, T and ϕ, respectively. The schematic representation of the contact behaviour with the implemented CWFS approach in the generated GBMs is shown in Fig. 10.

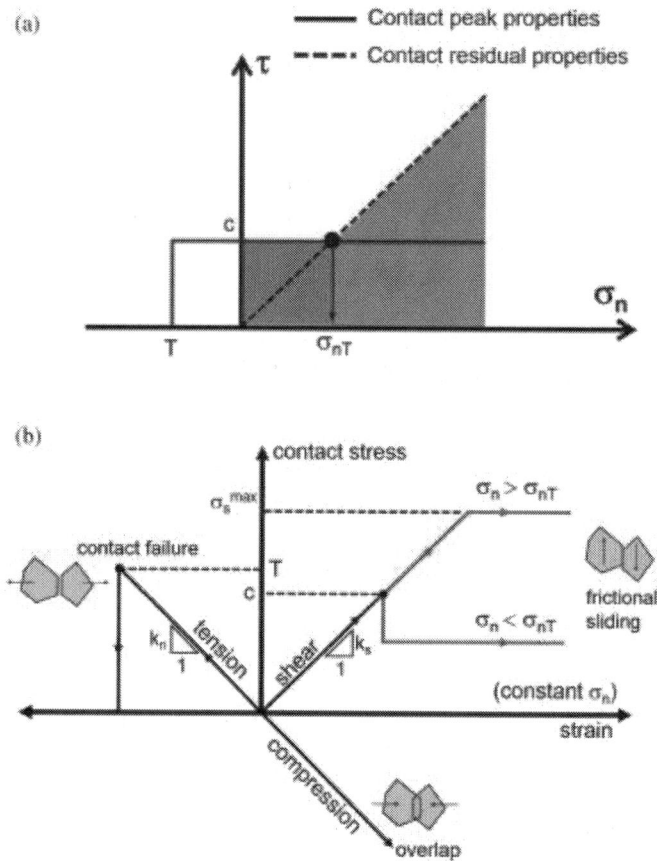

Figure 10. 3DEC simulation of fracturing along grain boundaries. (a) Sequential mobilization of cohesion and frictional component of contact shear strength. (b) Constitutive behaviour of grain contacts with implemented CWFS approach (σ_{nT} is the normal stress at which the contact properties are switched from peak to residual, σ_{smax} is the maximum shear stress at a contact with normal stress greater than σ_{nT}).

For the sensitivity analysis of the GBM response to the grain contact properties, a set of micro-parameters were chosen to closely replicate the mechanical properties of the Lac du Bonnet granite (Diederichs, 2000). The micro-properties were adjusted accordingly in an iterative process suggested by Potyondy and Cundall (2004) and Itasca (2008) until the desired macro-response was reached in terms of UCS and other macro-properties. The final micro-properties, model properties and stress–strain curve for the model are shown in Table 1 and Fig. 11, respectively. The micro-properties listed in Table 1 will be used in Section 5 for the parametric study.

Table 1. Micro-parameters used for calibration of the 3D Voronoi GBM and the resulting model macro-response.

Contact micro-parameters										Model macro-properties			
k_n(GPa/m	k_n/k_s	c_p(MPa)	c_r(MPa)	T_p(MPa)	T_r(MPa)	ϕ_p(°)	ϕ_r(°)	E (GPa)	v	UCS (MPa)	CI (MPa)	E (GPa)	v
68,000	5	148	0	14.8	0	0	25	100	0.2	183	111	64.7	0.3

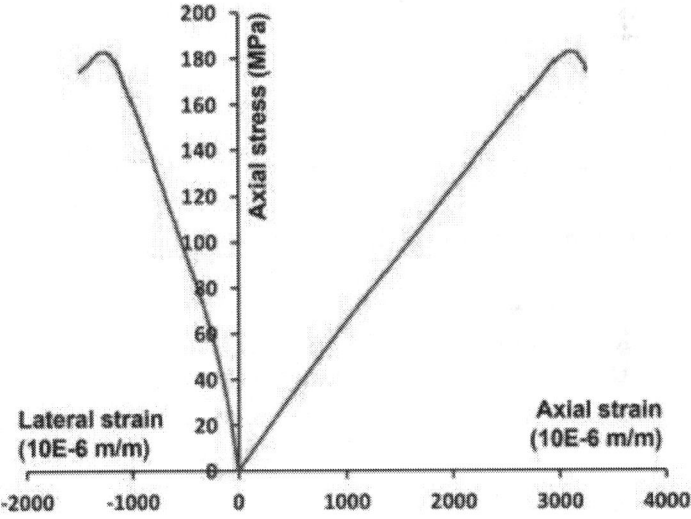

Figure 11. Stress-strain curve for the UCS model with the micro-parameters listed in Table 1.

SENSITIVITY ANALYSIS FOR GBM MICRO-PROPERTIES

The primary requirement prior to any GBM modelling is to calibrate the model micro-properties. The calibration process can be challenging and laborious if done in a random fashion. Therefore, a series of sensitivity analyses were performed on the GBM contact properties and grain size to better understand the role of these micro-properties on the overall response of the model and to establish the calibration procedure for a 3D Voronoi GBM. The sensitivity of the model response to the contact micro-properties was investigated by varying one micro-property at a time while the rest of the micro-parameters were kept constant, equal to the values listed in Table 1. The grain elastic constants (Young's modulus and Poisson's ratio) were considered real and were obtained from reported laboratory tests on the composing minerals of the Lac du Bonnet granite (Lama and Vutukuri, 1978) as opposed to rigid (Kazerani and Zhao, 2010), or semi-rigid (by assuming very high stiffnesses for them (Itasca, 2011)).

Contact normal stiffness (k_n)

The earlier work by Kazerani and Zhao (2010) on 2D rigid Voronoi GBMs and similar work by Gao and Stead (2014) on UDEC (Itasca, 2013b) trigon models, suggested that the simulated Young's modulus of the specimen is strongly correlated with the normal and shear stiffnesses of the contacts between the grains. They showed that by increasing the normal and shear stiffnesses of the contacts, the model's Young's modulus increases. Similarly, the contact stiffness in the 3D Voronoi models directly controls the specimen's Young's modulus as shown in Fig. 12. The increase in the normal and shear stiffnesses of the contacts stiffens the model's contact lattice and consequently enhances the stiffness of the model. Stiffening the models contact lattice also increases the role of grain deformation in controlling the model's macro-behaviour. Therefore, increasing the contact normal and shear stiffness values changes the models Poisson's ratio to within a closer range to the grain's Poisson's ratio. For the models shown in Fig. 12, the models overall Poisson's ratio approaches 0.20 (grains Poisson's ratio) with increasing contact stiffness values. The Poisson's ratio of the model with constant k_n/k_s ratio is independent of the change in the contact stiffness when grains are assumed rigid (Kazerani and Zhao, 2010).

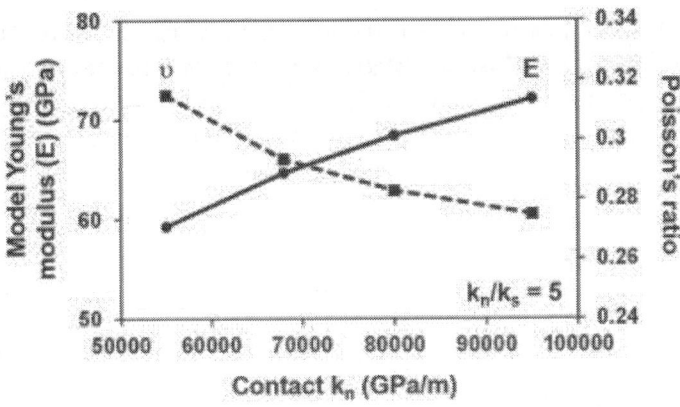

Figure 12. Effect of contact normal stiffness on the model's elastic constants.

Contact normal to shear stiffness ratio (k_n/k_s)

The Poisson's ratio is directly governed by the normal to shear stiffness ratio of the grain bonds in the BPM formulation as demonstrated by Diederichs, 2000 and Potyondy and Cundall, 2004 and Cho et al. (2007). Similar works in UDEC with rigid Voronoi grains (Kazerani and Zhao, 2010) and elastic trigon grains (Gao and Stead, 2014) support this.

The stress–strain curves from the sensitivity analysis for the contact k_n/k_s ratio for the 3D Voronoi GBM is shown in Fig. 13. In this series of models, the contact normal stiffness (k_n) was kept constant, at 68,000 GPa/m, while different k_n/k_s ratios (2, 5, 8 and 16) were examined. The change in the model's elastic constants as a function of k_n/k_s ratio is plotted in Fig. 14. This figure shows that increasing the k_n/k_s ratio increases the Poisson's ratio, even beyond the realistic threshold of 0.5, before the crack initiation threshold is reached. Conversely, increasing the k_n/k_s ratio decreases the Young's modulus and reduces the brittleness (increases ductility) of the model. By decreasing the shear stiffness and keeping the normal stiffness constant (increasing k_n/k_s ratio), while this makes the model softer and consequently lowers the Young's modulus, the contacts are encouraged to dilate in shear as oppose to extensile opening, which results in a larger lateral deformation and thus a larger Poisson's ratio. Therefore in 3D Voronoi GBMs, the k_n/k_s ratio is not only the major controlling factor for the Poisson's ratio, but also controls the Young's modulus and the brittle to ductile transition of the model (Fig. 13).

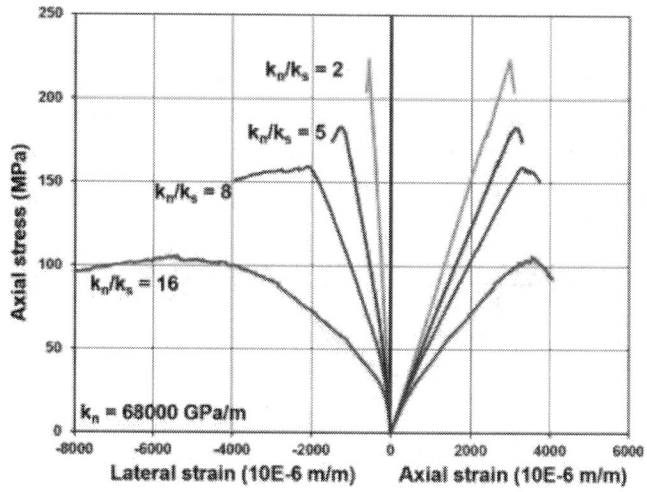

Figure 13. Stress-strain curves for the models with constant normal stiffness (68,000 GPa/m) but different ratios of contact normal to shear stiffness.

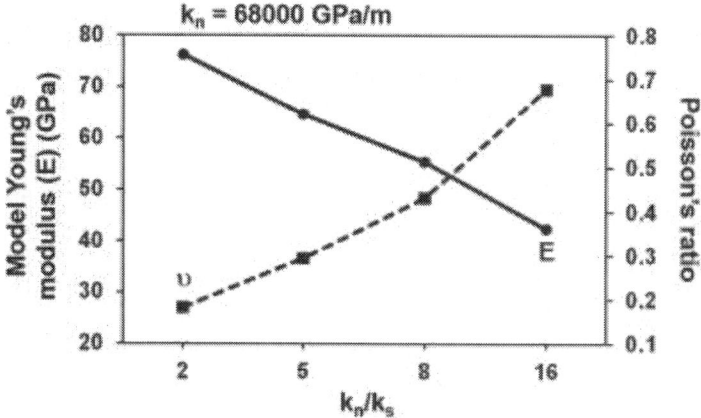

Figure 14. Effect of contact normal to shear stiffness ratio (k_n/k_s) on the model's elastic constants.

Contact c/T

The contact or bond strength properties in UDEC and PFC directly control the strength of the models in compression or tension. Increasing the contact strength-related micro-properties improves the specimen's strength (Diederichs, 2000, Potyondy and Cundall, 2004 and Gao and Stead, 2014). In Diederichs (2000), the shear to normal strength ratio of the contact bonds in PFC controls the peak strength and brittleness of the model. The material response shifts towards a more brittle fashion with increasing shear to tensile bond strength ratio. This is because increasing this ratio for the contact bonds forces the micro-fractures to form mostly in tension, which is similar to the initiation of extensile fractures in laboratory brittle rock specimens (Diederichs, 2007).

A suite of models were run with identical grain contact peak cohesion (c), but varying peak tensile strengths (T) for the grain contacts to achieve different ratios of cohesion to tensile strengths (c/T) for the 3D Voronoi GBMs. The UCS and UCS/CI results for this series of models are shown in Fig. 15. As expected, the peak strength of the specimen increases with decreasing c/T ratio, since T is increasing, causing the strength of the grain contacts to improve. The other macro-property that is affected by the c/T ratio is the CI threshold in the models. The opening of fractures at the CI threshold is assumed to be in an extensile fashion and therefore governed by the tensile strength of the contacts (Diederichs, 2003). This is supported by the models in Fig. 15 which shows that by decreasing the c/T ratio, the UCS/CI ratio decreases drastically (increasing CI with increasing contact tensile strength).

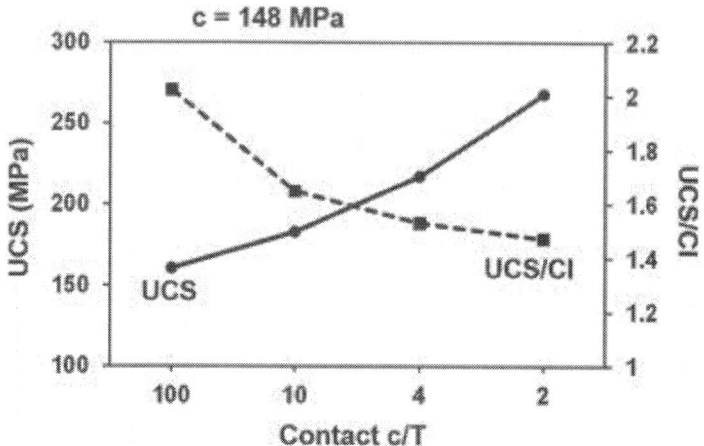

Figure 15. Effect of contact c/T ratio on the model's peak strength and UCS/CI ratio.

Diederichs, 2000 and Diederichs, 2003 reported limited success with PFC2D for simulation of crack propagation beyond a single disk at the CD threshold. This is due to stress redistribution amongst the bonds in the vicinity of a bond once it breaks, which impedes stress concentration at the crack tips and therefore suppresses crack extension. Gao and Stead (2014) argued that they have successfully simulated the crack propagation and coalescence within their trigon approach in UDEC and 3DEC. This was however solely based on the visual evaluation of contact openings. Martin and Chandler (1994) identified the deviatoric stress corresponding to the maximum volumetric strain of a laboratory specimen under compressive loading, as the onset of unstable cracking, or in other words, the CD threshold. Here, the volumetric strain is plotted versus the axial stress normalized to the peak strength of each specimen for models with different c/T ratios in Fig. 16. If the reversal of volumetric strain can also be identified as the crack propagation threshold in the numerical models, it can be seen that unstable micro-cracking or crack propagation occurs in models with c/T ratios equal to 4, 10 and 100. In Fig. 16, the decreasing c/T ratio shifts the axial stress associated with the reversal point of the volumetric strain closer to the UCS. For $c/T = 2$, the volumetric strain of the model never reaches the reversal point. Trans-granular cracks are major contributors to unstable cracking and the crack propagation threshold in brittle rocks. The absence of this factor can suppress the unstable cracking to a large extent in GBMs with unbreakable grains. Sub-tessellation of the Voronoi grains as described in Section 3.3allows for trans-granular fracture propagation and can be a solution to this limitation. The decreasing c/T ratio as observed in Fig. 16 increases the slope of the volumetric strain curve for the models. This can be attributed to the increase in the tensile strength of the contacts which allows for larger contact dilation before failure and therefore larger volumetric strains at the failure stress.

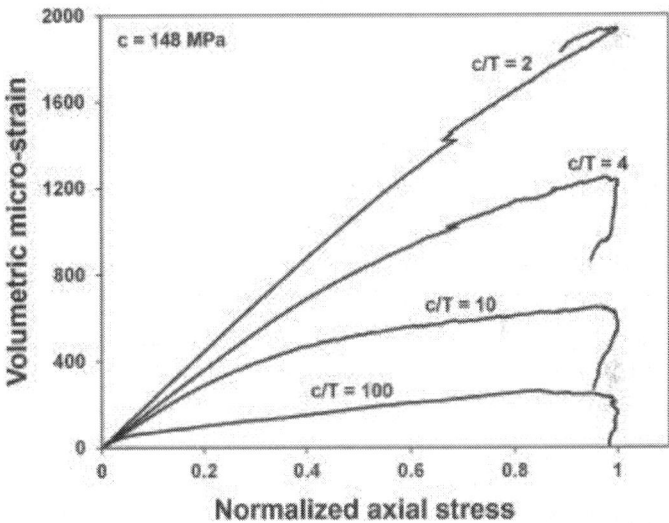

Figure 16. Effect of contact c/T ratio on the model's volumetric strain behaviour.

The cohesion, tensile strength and friction angle of the grain contacts are also shown in 2D Voronoi simulations of triaxial tests by Kazerani and Zhao (2010) and Gao and Stead (2014), to control the macro cohesion, tensile strength and friction angle of the model respectively.

Contact friction angle (ϕ)

To more accurately simulate the behaviour of brittle rocks, the CWFS approach was employed in the models and was discussed in detail in Section 4.4. The peak friction angle of the grain contacts was set to zero, therefore the material friction is controlled by the residual friction angle. This approach is more common in PFC modelling (Diederichs, 2003). According to Kazerani and Zhao (2010) and Gao and Stead (2014), the material friction angle can be controlled by the peak friction angle. In the CWFS approach, the material friction angle is governed by the residual friction angle. The results of a series of 3D Voronoi GBMs with different residual friction angles showed that, as expected, increasing the residual friction angle improves the strength of the models.

Effect of grain size

A comprehensive investigation on the effect of particle size on the PFC^{2D} and PFC^{3D}model macro-properties was done by Potyondy and Cundall (2004). They showed that in the PFC^{2D} models, the elastic constants and the peak strengths are independent of the particle size while in PFC^{3D}, the peak strength and the Young's modulus of the model are governed by the particle size. Similar to PFC^{2D}, the Poisson's ratios of PFC^{3D} models are independent of the particle size.

To investigate the effect of GBM grain size on the model response, a set of cylindrical specimens were carved from three cubes filled with 20,000, 40,000 and 60,000 Voronoi grains. The resulting specimens contained approximately 3000, 5300 and 7800 grains, with average grain volumes of 0.111 cm^3, 0.062 cm^3 and 0.042 cm^3, respectively. The stress–strain curves obtained from the UCS simulation for these models are shown inFig. 17. The results show that the peak strength of the models is independent of the grain size, within the grain sizes investigated in these models. A minor dependency of the model's elastic response to the grain size is observed to occur after the initiation of damage in the models (deviation of lateral strain from linearity (Martin, 1993)). As fracturing initiates, the macro-response of the models becomes slightly softer with decreasing grain size. This can be attributed to the relative magnitude of the grains elastic constants in comparison to the shear and normal stiffnesses of the grain contacts.

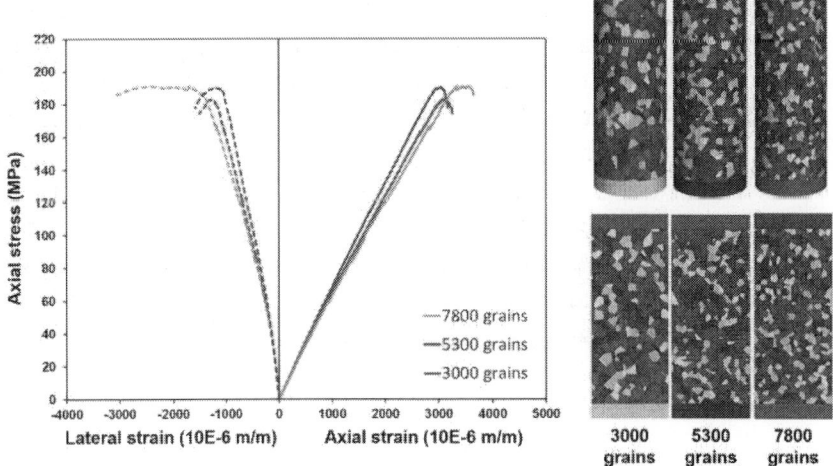

Figure 17. Effect of grain size on the material response.

CALIBRATION PROCEDURE FOR GBMS WITH REGULAR 3D VORONOI GRAINS

A better understanding of the role of each of the micro-parameter on the overall mechanical response of the GBMs was achieved from the sensitivity analyses. Therefore, the following procedure can be suggested for calibration of 3D Voronoi grain-based models from UCS data:

1) Decide on the Voronoi grain size (number of Voronoi cells within the specimen) based on the rock mineralogy and available computational power.
2) Define the elastic constants (Young's modulus and Poisson's ratio) for the grains according to laboratory testing or available literature for the composing minerals of the simulated rock.
3) Set the grain contact strength related components to high values or run the 3DEC models in "small strain" mode for calibration of the model elastic constants.

4) Adjust the contact normal to shear stiffness ratio (k_n/k_s) for the correct Poisson's ratio.

5) Adjust the normal stiffness (k_n) for the model to achieve a proper Young's modulus.

6) Cohesion, c/T and the residual friction angle of the grain contacts are adjusted in the model to achieve the correct peak strength and crack initiation threshold. In the case of matching the model response to triaxial and tensile strength testing data, the model cohesion, friction angle and tensile strength can be controlled by the grain contact cohesion, residual friction angle and tensile strength, respectively.

7) A distribution of strengths can be assigned to the grain contacts to further refine the crack initiation threshold.

8) Steps 3–6 (without changing the contact strength parameters to high values) can be reiterated for a more accurate calibration.

The suggested calibration procedure is for GBMs with elastic grains and grain contact behaviour as shown in Fig. 10.

DEM MODELLING OF DAMAGE PROGRESSION IN ANISOTROPIC ROCKS

Many rocks near the Earth's surface represent anisotropic behaviour due to layering, foliation, fissuring, bedding, stratification, jointing, etc. (Amadei, 1996). The available fabric arrangement in rocks directly influences their anisotropic behaviour (Gatelier et al., 2002). The term "fabric" is used in this paper as a general term to describe any planar feature in intact rock. The available fabric in intact rocks can be represented by one or a combination of the fabric elements shown in Fig. 18. The difference in mineral composition (Fig. 18a), preferred orientation of elongated grains (Fig. 18b, c), spatial variation of grain sizes (Fig. 18d), preferred orientation of platy minerals, aggregates or planar micro-fractures (Fig. 18e–g) or the existence of any combination of these elements in intact rock (Fig. 18h) can introduce anisotropic effects to its mechanical behaviour.

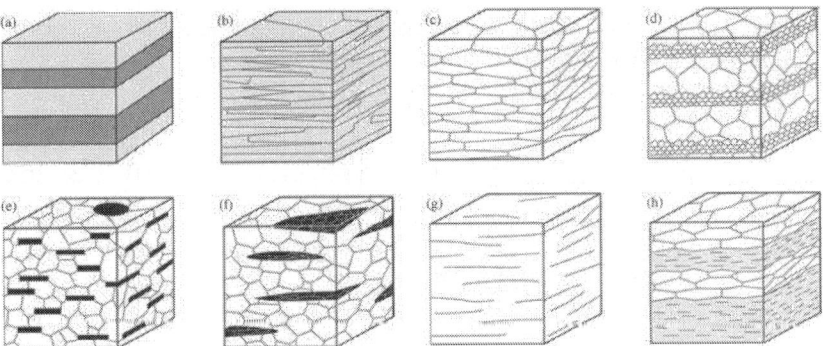

Figure 18. Fabric elements that can introduce anisotropic behaviour in intact rocks (after Passchier and Trouw (2005), modified from Hobbs et al. (1976)).

Previous attempts for explicit simulation of mechanical anisotropic behaviour for brittle intact rocks by different researchers have mainly concentrated on introducing anisotropic damage to the intact rock, similar to the fabric element as shown in Fig. 18g. Diederichs et al. (2004) studied the effect of pre-existing damage in brittle rocks on their mechanical behaviour such as CI and CD by inducing different crack intensities (removing or breaking bonds between disks) at different orientations in their PFC2D models. In a hybrid continuum/discontinuum approach, Lisjak et al. (2014a) successfully applied a small scale Discrete Fracture Network (DFN) to a combined FDEM model to study the anisotropic behaviour of Opalinus Clay. They later extended this approach to a larger scale to study the development of the Excavation Damaged Zone (EDZ) around tunnels excavated in this formation (Lisjak et al., 2014b).

The Voronoi tessellation provides an opportunity to generate GBMs that resemble the fabric elements similar to the ones in Fig. 18c–f at the micro-scale. In the simplest form, a GBM with different scaling factors can be generated from a regular Voronoi tessellation to mimic the intensity of the anisotropy in the intact rock. Fig. 19a–c which host the fabric elements as shown in Fig. 18c can be approximated by scaled Voronoi grains with different length/width ratios.

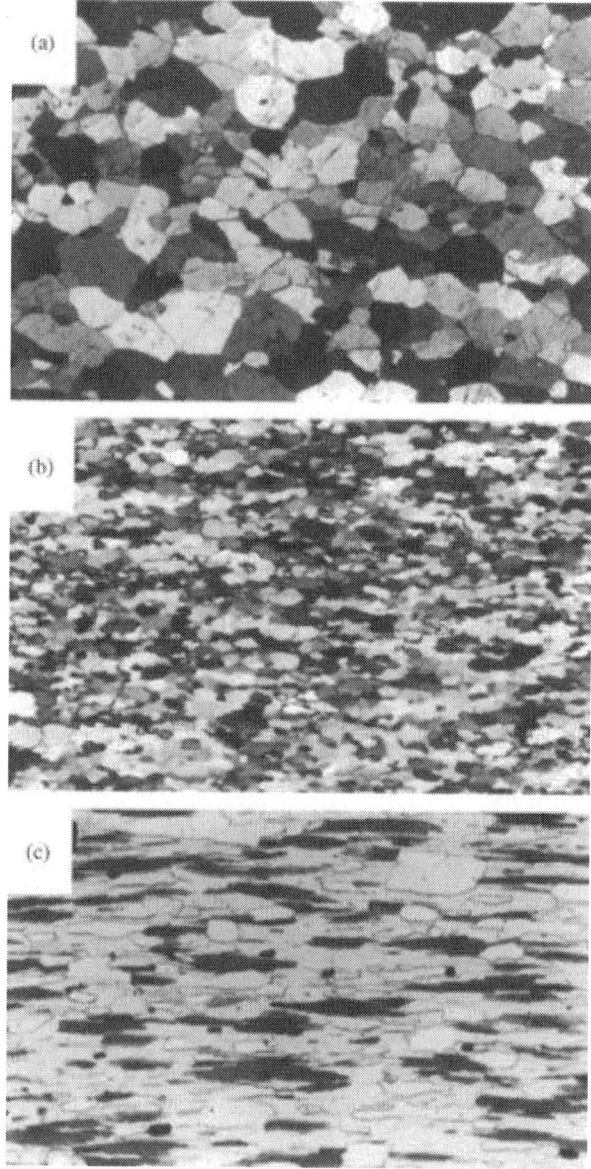

Figure 19. Crystalline minerals with various degrees of recrystallization. (a) Polygonal fabric of scapolite (width of view 4 mm). (b) Elongated grains in recrystallized quartz (width of view 1.8 mm). (c) Strongly orientated mineral grains defined by parallel grains of biotite, muscovite and quartz (width of view 1.8 mm), (modified after Passchier and Trouw (2005)).

Laminated GBM for fabric-guided micro-fracturing simulation
Laminated GBMs with fabrics composed of elongated 3D Voronoi grains can be constructed with scaling. An example for generation of a cylindrical laminated specimen oriented 45° with respect to the loading direction is illustrated in Fig. 20. A prismatic domain with height/width ratio equal to the desired length/width ratio of grains is generated, scaled and then rotated for proper fabric orientation. The generated model is then exported to 3DEC and cut for a cylindrical specimen.

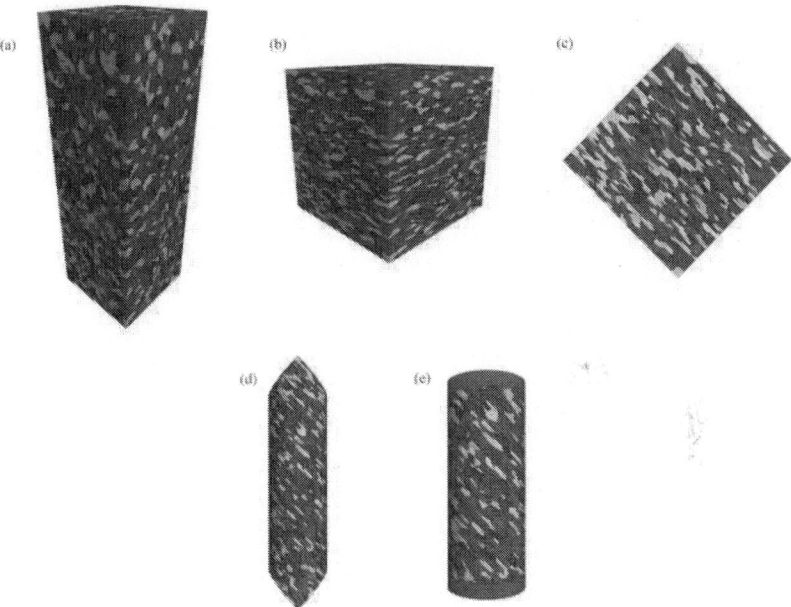

Figure 20. Generation of layered GBMs for anisotropic simulations in 3DEC (an example of a 45° specimen). (a) Generation of 3D Voronoi tessellations with a prismatic domain. (b) Scaling the domain to achieve the layered model with elongated grains. (c) Rotation of the domain to achieve the desired layering orientation. (d, e) Cutting through the block to make the cylindrical specimen and platens.

Calibration
To investigate the application of the laminated GBMs for simulation of anisotropic rock behaviour, a series of UCS models were generated in 3DEC by using the elongated grains for different fabric orientations, following the procedure introduced in Section 7.1. The Cobourg

Limestone, which is composed of alternating layers of fossil rich packstone and argillaceous wisps and blebs, was chosen for the model verification. The mechanical properties of the Cobourg Limestone were obtained from Ghazvinian et al., 2013a and Ghazvinian et al., 2013b. In these studies, Cobourg Limestone specimens with different layering orientations with respect to the loading direction were drilled from large diameter cores obtained from a limestone quarry in Bowmanville, Ontario. For five groups of specimens with apparent bedding oriented at 0°, 30°, 45°, 60° and 90°, four to six specimens were tested for their mechanical properties including the peak strength, elastic constants and crack damage thresholds. An example of the Cobourg Limestone specimens and the associated GBMs are shown in Fig. 21.

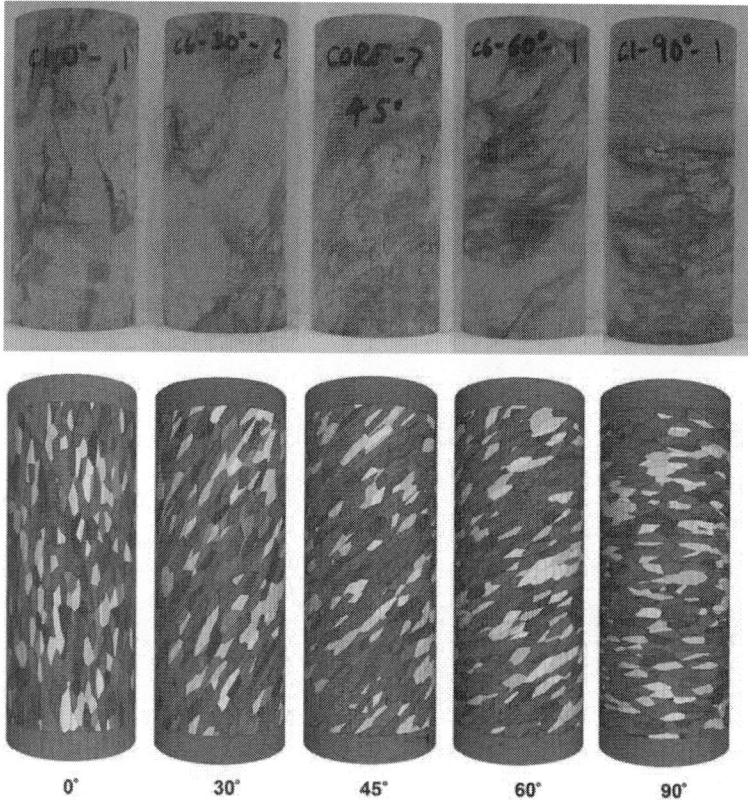

Figure 21. The Cobourg Limestone specimens with different fabric orientations with respect to the loading direction (before testing) and the corresponding 3DEC instances.

In addition to the micro-parameters discussed in Section 4.4 that need to be calibrated for GBMs with regular Voronoi tessellations, here the length to width (l/w) ratio of the grains for laminated GBMs also needs to be calibrated. For a specific rock, the models with different fabric orientations have identical micro-parameters. It is the l/w ratio of the grains which defines the mechanical anisotropy for the behaviour of the GBMs with different fabric orientations.

The calibration process was conducted by deciding on an initial value for the grains l/w ratio and adjusting the micro-parameters for the horizontally (90°) and vertically (0°) laminated models simultaneously with an identical set of micro-parameters to achieve the proper elastic constants. Then, the strength-related micro-parameters were adjusted for the horizontal model to calibrate its strength properties. Subsequently by applying those micro-parameters to the vertically laminated model and fine-tuning the grains l/w ratio to attain the proper strength macro-properties, the global set of micro-parameters for the entire suite of specimens with different fabric orientations was obtained. The stress–strain curves for the numerical and laboratory specimens with vertical and horizontal fabrics are shown in Fig. 22. The calibrated micro-properties, listed in Table 2, were applied to the suite of GBMs with five different fabric orientations as shown in Fig. 21.

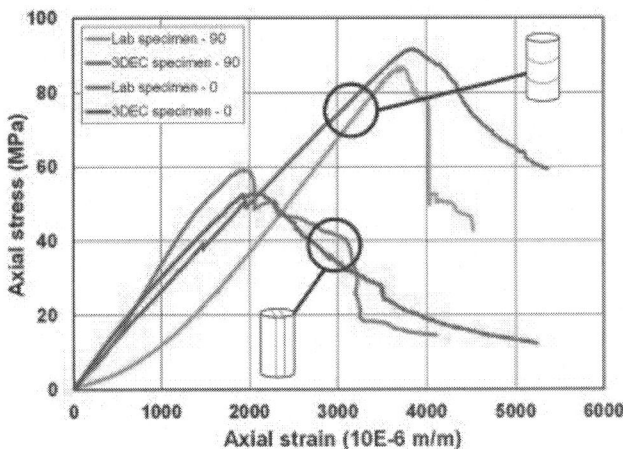

Figure 22. Stress-strain curves for 0° and 90° specimens tested in the laboratory and numerically simulated in 3DEC.

Table 2. Calibrated micro-parameters based on the horizontal and vertical laminated models.

k_n (GPa/m)	k_n/k_s	c_p (MPa)	c_r (MPa)	T_p (MPa)	T_r (MPa)	ϕ_p (°)	ϕ_r (°)	E (GPa)	v	l/w
46,000	8	50	0	5	0	0	25	50	0.1	2

Numerical results

The laminated GBM approach was successful in capturing the effect of fabric orientation on the failure mode of the models. In Fig. 23, which shows the fractures in the models after failure on a plane cut through the centre of the model, the switch between the failure modes can be observed to occur as a function of the fabric orientation. Axial splitting and brittle failure can be seen in the 0° and 90° specimens respectively and the formation of shear bands is evident in the other orientations. Remarkably the formation of axial micro-fractures and their role towards formation of a shear band by kinking can be seen in the 45° model.

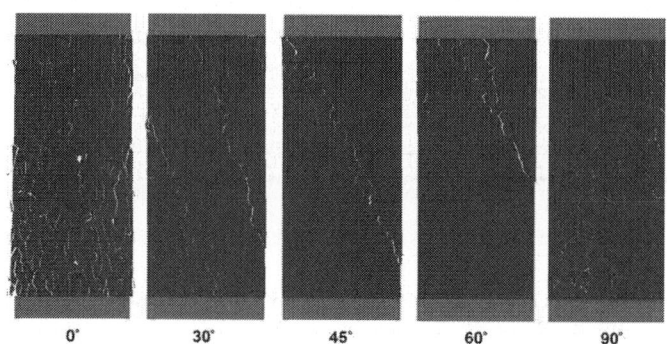

Figure 23. Cross-sections through the laminated GBMs after failure.

The comparison between the laboratory testing results of the Cobourg Limestone and the corresponding numerical macro-properties are shown in Fig. 24. The UCS and CI thresholds between the two sets of data are in good agreement and the classical U-shape behaviour (Jaeger and Cook, 1969) for anisotropic rocks is well-captured by the GBMs. The elastic response of the numerical models, which were calibrated before the strength properties, approximately captures the lower-bound Young's moduli (except the vertically and horizontally laminated models) and the upper-bound of the laboratory measured Poisson's ratios. Since calibration of the strength-related components of the macro-behaviour for GBMs takes place after the calibration of the elastic response, the Young's moduli and the Poisson's ratio of the models can be slightly altered, as observed in Fig. 24. Therefore, following the calibration procedure for a couple of iterations between adjustments for the elastic response and strength of the models can increase the accuracy of the calibrated micro-parameters.

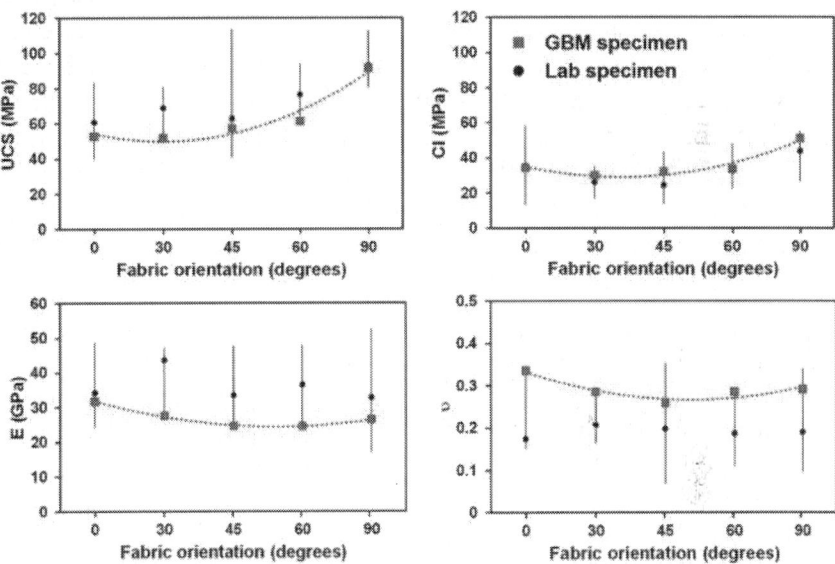

Figure 24. Comparison between the mechanical properties of the Cobourg Limestone measured in the laboratory and simulated in 3DEC (the dotted lines are 2nd order polynomial best fit to the 3DEC results).

Sensitivity analysis

The micro-properties that can be calibrated for a specific set of macro-properties in general are non-unique. A series of sensitivity analyses were conducted for the most important micro-properties of a laminated GBM to better understand their role in defining the classical U-shape behaviour that is anticipated from anisotropic rocks. The U-shape behaviour was first introduced by Jaeger and Cook (1969) to show the effect of discontinuities on the rock mass strength. It is employed here to demonstrate the variation of mechanical properties with respect to the intact rocks fabric orientation.

Scaling (grain *l/w* ratio)

Numerous models with different grain l/w ratios were achieved by scaling prismatic domains of different heights, but all filled with an identical number of grains, down to a proper cube. This was discussed in detail in Section 7.1. Two new sets of laminated GBMs with l/w ratios of 1.5 and 2.5 were generated and numerically tested by using the micro-properties listed in Table 2. The results for these two new sets of models are compared in Fig. 25 with the original suite of GBMs as discussed in Section 7.3.

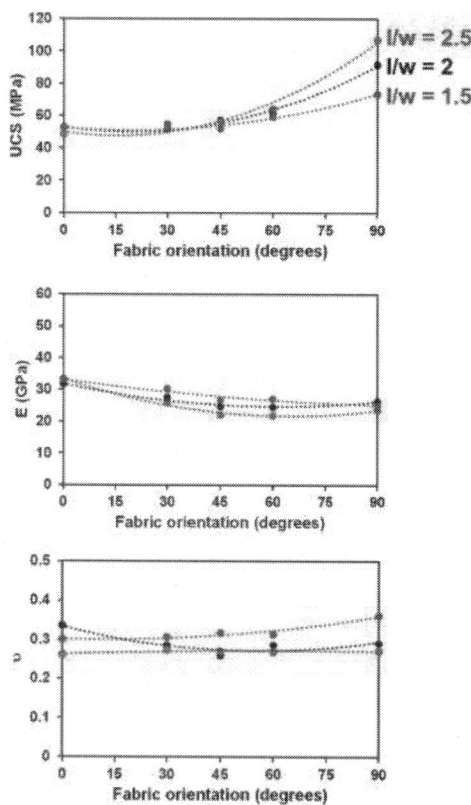

Figure 25. Effect of grain l/w ratios on the GBM macro-response (the dotted lines are 2nd order polynomial best fit to the data points).

The l/w ratio of grains appears to have a major impact on the strength of the horizontally laminated models, while its effect on the strength of models with other fabric orientations seems to be minute. Regarding the Young's modulus and Poisson's ratio, the l/w ratio appears to control the depth of the anisotropy U-shape curve, and the elevation of the curve.

Contact c/T ratio

The c/T ratio of contacts in GBM plays a key role in controlling the macro-response of the models as observed in Section 5.3. Two new sets of models with identical micro-parameters to the original set of laminated models (as listed in Table 2, $c/T = 10$), but different contact tensile strengths (T), to achieve c/T ratios of 2 and 20 were numerically tested. The results are plotted in Fig. 26. Similar to the GBM with regular Voronoi grains, reducing the c/T ratio by increasing the tensile strength (T) of contacts improves the peak strength of the models and subsequently shifts the U-shape anisotropy curve to a higher elevation (model with higher peak strengths). This is also true for the effect of the c/T ratio on the Young's moduli of the models. The c/T ratio controls the flatness and the direction of the anisotropy curve for the CI thresholds and Poisson's ratios as shown in Fig. 26. The c/T ratio is particularly important for its role in controlling the anisotropy curve for the CI threshold. In some rocks, the CI threshold is entirely independent of the fabric orientation (Hakala et al., 2007) while it can change according to the orientation of the lamination in other rock types (Ghazvinian et al., 2013a). Therefore the c/T parameter can be used to adjust the sensitivity of the CI threshold to the orientation of the fabric present in a rock.

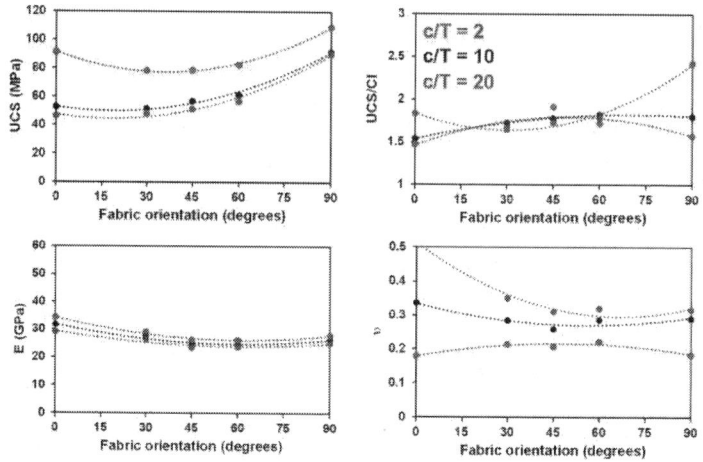

Figure 26. Effect of contact c/T ratio on the mechanical behaviours of laminated GBMs (the dotted lines are 2nd order polynomial best fit to the data points).

Contact friction angle

Increasing the friction angle of the contacts (i.e. the residual friction angle according to the contact model as shown in Fig. 10) improves the peak strengths of the models. The increase in the strength of the models intensifies as the orientation of fabric in the rock changes from vertical to horizontal. As shown in Fig. 27, the maximum strength increase of the laminated GBMs as a function of increasing the models' friction angle is observed in the horizontally laminated model.

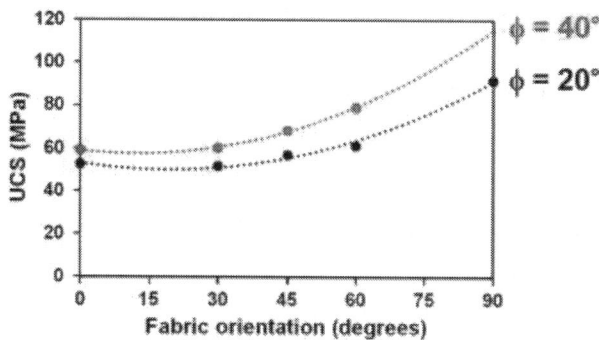

Figure 27. Effect of contact friction angle on the strength of laminated GBMs (the dotted lines are 2nd order polynomial best fit to the data points).

Calibration procedure

Verification of the laminated GBM approach with the available test data from the Cobourg Limestone confirmed the application of this modelling method for numerical replication of brittle rock behaviour with inherent anisotropic mechanical properties. Based on the conducted sensitivity analyses for micro-properties, the calibration procedure for the GBMs with elongated 3D Voronoi grains can be summarised as follows:

1) Decide on the grain size for the GBM and an initial grain l/w ratio.
2) Define the elastic constants (Young's modulus and Poisson's ratio) for the grains. It is nearly impossible to match the elastic constants for the horizontally and vertically laminated models during the calibration stage with an identical set of micro-parameters with rigid grains or unrealistically very stiff grains.
3) Set the strength components of the contacts to a high value so that the deformational behaviour of the model is fully elastic (or set the "small strain" in 3DEC) and adjust the Poisson's ratio for the models with the horizontal and vertical fabric simultaneously by modifying the k_n/k_s ratio.
4) Calibrate the Young's moduli for the models with the horizontal and vertical fabric simultaneously by adjusting the k_n for the contacts.
5) Calibrate the peak strength and CI threshold for the horizontally laminated model by fine-tuning the c/T ratio and c. Adjust the friction angle based on the post-peak behaviour (if needed).
6) Apply the micro-parameters calibrated thus far to the vertically laminated model and adjust the grains l/w ratio accordingly.
7) Apply the micro-parameters calibrated thus far to the entire suite of laminated GBMs and correct the U-shape anisotropy curve by adjusting the micro-parameters according to the discussions in Section 7.4 for a few iterations.

DISCUSSION

This paper showed that brittle rock damage through explicit simulation of micro-fracturing can be accurately captured in numerical models by means of GBMs built with 3D Voronoi grains. Similar work has been done previously in 2D (e.g. Kazerani and Zhao, 2010, Lan et al., 2010 and Itasca, 2011), however the complexity of generating proper 3D Voronoi tessellations has encumbered its development in 3D DEM codes to this point. The mechanics of the Voronoi tessellated models at the micro-scale remains nearly the same in transition from 2D to 3D. However some complexities in regards to the calibration process can intensify when switching from UDEC to 3DEC for Voronoi GBMs. In UDEC and 3DEC, the contact models commonly used for grains are the constitutive slip models developed for rock mass joints and therefore, attempting to formulate the micro-scale grain contact behaviour by using an analogue established for large-scale discontinuities can be challenging. In this study, the CWFS approach was successfully implemented in the grain contact model to better address the peak and residual friction angles and gradual mobilization of the frictional strength component of the contacts. This approach can be further refined with GBM simulations under triaxial conditions to capture the correlation between the material friction angle and the friction angle in the CWFS model for grain contacts.

Grain contact normal and shear stiffnesses are two of the most important micro-parameters in terms of defining the model's macro-response. The transition from 2D to 3D models drastically increases the importance of these two micro-parameters. The unrealistic representation of contact stiffnesses (particularly the normal stiffness) with a constant value leads to high stiffness parameters for the contacts and therefore an enormous stiffness contrast between the grain and the grain contact, which can be erroneous. It is anticipated that defining the normal stiffness of the grain contact as a function of the grains' overlapping area (in 2D) or volume (in 3D) can better represent the micro-mechanical behaviour of crystalline rocks.

3D Voronoi tessellated models can be used as the building block for the Synthetic Rock Mass (SRM) (Mas Ivars et al., 2011) to represent intact rock. Its combination with the in-situ joint network, commonly referred to as the DFN, can represent rock masses which can be numerically investigated. The regular and laminated GBMs can be used for SRM representation of rock masses, including intact rock with isotropic or inherent anisotropic mechanical behaviour. Success of the 3D Voronoi GBMs in simulating the micro-fracturing in laboratory scale specimens supports its ability in simulation of rock damage and fracturing in the walls of deep underground excavations in brittle rocks by means of proper up-scaling rules.

CONCLUSIONS

The concerns with the formulation of bonded particle models (i.e. PFC) lead to increasing application of Voronoi grain-based models for simulation of fracture formation in rocks despite its poor computational speed compared to the BPM. While clumping and clustering in the BPM (Cho et al., 2007 and Ghazvinian, 2010) and the modifications to the BPM formulation (development of the flat-joint contact model (Potyondy, 2012)) in the recent years have solved the concerns with the accuracy of the BPM results to a great extent, the Voronoi based GBMs are still popular due to simplicity in generating these models. To extend the application of the Voronoi tessellated DEM models from 2D to 3D, a new methodology was introduced in this paper for generation of models with any arbitrary 3D convex shape domains in 3DEC tessellated with Voronoi grains. Furthermore, a new model was introduced for simulation of inherent mechanical anisotropic behaviour of rocks containing a fabric based on the freshly developed 3D Voronoi GBM. The fabric-guided micro-fracturing with this approach was verified successfully with the available laboratory testing data of the Cobourg Limestone. The sensitivity analysis of the grain contact micro-properties proved that the dependency or independency of the CI threshold to the fabric orientation in the rock can be controlled for different rocks and different types of anisotropic behaviours and finally was employed to establish a calibration procedure for the 3D Voronoi laminated grain-based models.

CONFLICT OF INTEREST

The authors wish to confirm that there are no known conflicts of interest associated with this publication and there has been no significant financial support for this work that could have influenced its outcome.

ACKNOWLEDGEMENTS

The authors would like to acknowledge the Nuclear Waste Management Organization of Canada (NWMO), the Swedish Nuclear Fuel and Waste Management Company (SKB) and the National Science and Engineering Research Council of Canada (NSERC) for supporting this research. The discussions with Dr. Jim Hazzard and Dr. Branko Damjanac from the Itasca Consulting Group significantly helped in the preparation of this paper. Special thanks to Felipe Duran, Michelle van der Pouw Kraan and Dr. Derek Martin for their inputs. Assistance received from Mark Jensen and Tom Lam from NWMO, Rolf Christiansson from SKB and Denis Labrie from Natural Resources Canada is much appreciated.

REFERENCES

1. Amadei B. Importance of anisotropy when estimating and measuring in situ stresses in rock. International Journal of Rock Mechanics and Mining Sciences and Geomechanics Abstracts 1996;33(3):293e325.
2. Barbe F, Quey R, Musienko A, Cailletaud G. Three-dimensional characterization of strain localization bands in high-resolution elastoplastic polycrystals. Mechanics Research Communications 2009;36(7):762e8.
3. Barbe F, Quey R. A numerical modelling of 3D polycrystal-to-polycrystal diffusive phase transformations involving crystal plasticity. International Journal of Plasticity 2011;27(6):823e40.

4. Benedetti I, Aliabadi MH. A three-dimensional cohesive-frictional grain-boundary micromechanical model for intergranular degradation and failure in polycrystalline materials. Computer Methods in Applied Mechanics and Engineering 2013;265:36e62.

5. Cho N, Martin CD, Sego DC. A clumped particle model for rock. International Journal of Rock Mechanics and Mining Sciences 2007;44(7):997e1010.

6. Cundall PA, Hart RD. Development of generalized 2-D and 3-D distinct element programs for modeling jointed rock. Minneapolis, MN, USA: Itasca Consulting Group Inc.; 1985.

7. Cundall PA. Formulation of a three-dimensional distinct element model d Part I. A scheme to detect and represent contacts in a system composed of many polyhedral blocks. International Journal of Rock Mechanics and Mining Sciences and Geomechanics Abstracts 1988;25(3):107e16.

8. Damjanac B, Board M, Lin M, Kicker D, Leem J. Mechanical degradation of emplacement drifts at Yucca Mountain d a modeling case study: part II: lithophysal rock. International Journal of Rock Mechanics and Mining Sciences 2007;44(3):368e99.

9. Diederichs MS. Instability of hard rock masses: the role of tensile damage and relaxation. PhD Thesis. Waterloo, Ontario, USA: University of Waterloo; 2000. Diederichs MS. Rock fracture and collapse under low confinement conditions. Rock Mechanics and Rock Engineering 2003;36(5):339e81.

10. Diederichs MS, Kaiser PK, Eberhardt E. Damage initiation and propagation in hard rock during tunnelling and the influence of near-face stress rotation. International Journal of Rock Mechanics and Mining Sciences 2004;41(5):785e812.

11. Diederichs MS. The 2003 Canadian geotechnical colloquium: mechanistic interpretation and practical application of damage and spalling prediction criteria for deep tunnelling. Canadian Geotechnical Journal 2007;44(9):1082e116.

12. Gao FQ, Stead D. The application of a modified Voronoi logic to brittle fracture modelling at the laboratory and field scale. International Journal of Rock Mechanics and Mining Sciences 2014;68:1e14.

13. Gatelier N, Pellet F, Loret B. Mechanical damage of an anisotropic porous rock in cyclic triaxial tests. International Journal of Rock Mechanics and Mining Sciences 2002;39(3):335e54.

14. Ghazvinian E, Diederichs MS, Martin CD. Identification of crack damage thresholds in crystalline rock. In: Proceedings of Eurock 2012, Stockholm, Sweden; 2012.

15. Ghazvinian E, Perras M, Diederichs M, Labrie D. The effect of anisotropy on crack damage thresholds in brittle rocks. In: Proceedings of the 47th US Rock Mechanics/Geomechanics Symposium. San Francisco, California, USA: American Rock Mechanics Association; 2013a.

16. Ghazvinian E, Perras M, Langford C, Diederichs MS, Labrie D. A comprehensive investigation of crack damage anisotropy in Cobourg limestone and its effect on the failure envelope. In: Proceedings of GeoMontreal 2013. Montreal, QC, Canada: Canadian Geotechnical Society; 2013b.

17. Ghazvinian E. Modelling and testing strategies for brittle fracture simulation in crystalline rock samples. MS Thesis. Kingston, Ontario: Queen's University; 2010.

18. Hajiabdolmajid V, Kaiser PK, Martin CD. Modeling brittle failure of rock. International Journal of Rock Mechanics and Mining Sciences 2002;39:731e41.

19. Hakala M, Kuula H, Hudson JA. Estimating the transversely isotropic elastic intact rock properties for in situ stress measurement data reduction: a case study of the Olkiluoto mica gneiss, Finland. International Journal of Rock Mechanics and Mining Sciences 2007;44(1):14e46.

20. Hobbs BE, Means WD, Williams PF. An outline of structural geology, vol. 570. New York: Wiley; 1976.

21. Itasca. 3DEC (3 dimensional distinct element code) version 5.0. Minneapolis, MN, USA: Itasca Consulting Group Inc.; 2013a.

22. Itasca. UDEC (Universal distinct element code). Minneapolis, MN, USA: Itasca Consulting Group Inc.; 2013b.

23. Itasca. FLAC (Fast Lagrangian analysis of Continua) version 7.0. Minneapolis, MN, USA: Itasca Consulting Group Inc.; 2012.

24. Itasca. Long-term geomechanical stability analysis: OPG's deep geological repository for low & intermediate level waste. 2011. Technical Report, NWMO DGR-TR-2011-17. Itasca. PFC3D (Particle flow code 3D) version 4.0. Minneapolis, MN, USA: Itasca Consulting Group Inc.; 2008.

25. Jaeger JC, Cook NG. Fundamentals of rock mechanics. London, UK: Methuen & Co Ltd.; 1969. Kanit T, Forest S, Galliet I, Mounoury V, Jeulin D. Determination of the size of the representative volume element for random composites: statistical and numerical approach. International Journal of Solids and Structures 2003;40:3647e 79.

26. Kazerani T, Zhao J. Micromechanical parameters in bonded particle method for modeling of brittle material failure. International Journal for Numerical and Analytical Methods in Geomechanics 2010;34(18):1877e95.

27. Kazerani T. A discontinuum-based model to simulate compressive and tensile failure in sedimentary rock. Journal of Rock Mechanics and Geotechnical Engineering 2013;5(5):378e88.

28. Lama RD, Vutukuri VS. Handbook on mechanical properties of rock, vol. II. Clausthal, Germany: Trans Tech Publications; 1978.

29. Lan H, Martin CD, Hu B. Effect of heterogeneity of brittle rock on micromechanical extensile behavior during compression loading. Journal of Geophysical Research 2010;115:1e14.

30. Lisjak A, Tatone BS, Grasselli G, Vietor T. Numerical modelling of the anisotropic mechanical behaviour of Opalinus Clay at the laboratory-scale using FEM/DEM. Rock Mechanics and Rock Engineering 2014a;47(1):187e206.

31. Lisjak A, Grasselli G, Vietor T. Continuum-discontinuum analysis of failure mechanisms around unsupported circular excavations in anisotropic clay shales. International Journal of Rock Mechanics and Mining Sciences 2014b;65:96e115.

32. Mahabadi OK, Lisjak A, Grasselli G, Munjiza A. Y-Geo: a new combined finitediscrete element numerical code for geomechanical applications. International Journal of Geomechanics 2012;12(6):676e88.

33. Martin CD, Chandler NA. The progressive fracture of Lac du Bonnet granite. International Journal of Rock Mechanics and Mining Sciences and Geomechanics Abstracts 1994;31(6):643e59.

34. Martin CD. Seventeenth Canadian geotechnical colloquium: the effect of cohesion loss and stress path on brittle rock strength. Canadian Geotechnical Journal 1997;34(5):698e725.

35. Martin CD. The strength of massive Lac du Bonnet granite around underground openings. PhD Thesis. Winnipeg, Manitoba: University of Manitoba; 1993.

36. Mas Ivars D, Pierce ME, Darcel C, Reyes-Montes J, Potyondy DO, Young RP, Cundall PA. The synthetic rock mass approach for jointed rock mass modelling. International Journal of Rock Mechanics and Mining Sciences 2011;48(2):219e 44.

37. Passchier CW, Trouw RAJ. Microtectonics. 2nd ed. Berlin: Springer; 2005

38. Potyondy DO, Cundall PA. A bonded-particle model for rock. International Journal of Rock Mechanics and Mining Sciences 2004;41(8):1329e64.

39. Potyondy DO. A flat-jointed bonded-particle material for hard rock. In: Proceedings of the 46th U.S. Rock Mechanics/Geomechanics Symposium. Chicago, USA: American Rock Mechanics Association; 2012.

40. Quey R, Dawson PR, Barbe F. Large-scale 3D random polycrystals for the finite element method: generation, meshing and remeshing. Computer Methods in Applied Mechanics and Engineering 2011;200(17):1729e45.

41. Quey R, Dawson PR, Driver JH. Grain orientation fragmentation in hot-deformed aluminium: experiment and simulation. Journal of the Mechanics and Physics of Solids 2012;60(3):509e24.

42. Quey R. Neper: numerical descriptors of polycrystals (version 2.0). 2014. Available at: http://neper.sourceforge.net/. Scholtès L, Donzé FV. A DEM model for soft and hard rocks: role of grain interlocking on strength. Journal of the Mechanics and Physics of Solids 2013;61(2): 352e69.

43. Yoon JS, Jeon S, Stephansson O, Zang A, Dresen G. A new method of microparameter determination for PFC2D synthetic rock model generation. In: Proceedings of the 1st International FLAC/DEM Symposium on Numerical Modeling. Minneapolis, MN, USA: Itasca Consulting Group Inc.; 2008.

44. You S, Zhao GF, Ji HG. Model for transversely isotropic materials based on distinct lattice spring model (DLSM). Journal of Computers 2011;6(6):1139e44.

CITATION

E. Ghazvinian, M.S. Diederichs, R. Quey, 3D random Voronoi grain-based models for simulation of brittle rock damage and fabric-guided micro-fracturing, Journal of Rock Mechanics and Geotechnical Engineering, Volume 6, Issue 6, December 2014, Pages 506-521, ISSN 1674-7755, http://dx.doi.org/10.1016/j.jrmge.2014.09.001.

CHAPTER 4

Analysis of Hydro-Mechanical Processes in a Ventilated Tunnel in an Argillaceous Rock on the Basis of Different Modelling Approaches

B. Garitte[1], A. Bond[2], A. Millard[3], C. Zhang[4], C. Mcdermott[5], S. Nakama[6], A. Gen[1]

[1] Universidad Politécnica de Catalunya, Barcelona, Spain

[2] Quintessa, Warrington, UK

[3] Commissariat à l'Energie Atomique, Gif-sur-Yvette, France

[4] Chinese Academy of Sciences, Wuhan, China

[5] ECOSSE, University of Edinburgh, Edinburgh, UK

[6] Japan Atomic Energy Agency, Tokai, Japan

ABSTRACT

In this paper, a modelling benchmark exercise from the DECOVALEX-2011 project is presented. The benchmark is based on the performance and results of a laboratory drying test and of the ventilation experiment (VE) carried out in the Mont Terri Underground Rock Laboratory (URL). Both tests involve Opalinus clay. The work aims at the identification, understanding and quantification of mechanisms taking place during the ventilation of a gallery in argillaceous host rocks on one hand and at investigating the capacity of different codes and individuals to reproduce these processes on the other hand. The 4-year in situ VE took place in a 1.3 m diameter unlined tunnel and included two resaturation–desaturation cycles. The test area was equipped with over one hundred sensors (including the global water mass balance of the system, relative humidity (RH), water content, liquid pressure, relative displacement and concentration of some chemical species) to monitor the rock behaviour during ventilation. The laboratory drying experiment, carried out before the VE, was designed to mimic the in situ conditions. The work was organized in a progressive manner in terms of complexity of the computations to be performed, geared towards the full hydro-mechano-chemical (HMC) understanding of the VE, the final objective. The main results from the modelling work reported herein are that the response of the host rock to ventilation in argillaceous rocks is mainly governed by hydraulic processes (advective Darcy flow and non-advective vapour diffusion) and that the hydro-mechanical (TM) back coupling is weak. A ventilation experiment may thus be regarded as a large scale-long time pump test and it is used to determine the hydraulic conductivity of the rock mass.

INTRODUCTION

During the construction and operation phases of a radioactive waste repository, the underground drifts will be subjected to a more or less intense ventilation period, which could produce partial desaturation of the rock around the drifts. Design of future storage facilities for radioactive waste in deep geological media requires a thorough understanding of the mechanisms occurring near the installations. This understanding is achieved by the parallel development of in situ tests, mimicking expected storage conditions such as mechanical loading, ventilation, heating, and theoretical thermo-hydro-mechanical-chemical (THMC) formulations. On one hand, the exploration of the THMC formulations allows the identification of possible processes involved near storage galleries. On the other hand, the analysis of in situ tests establishes the validity of the formulations based on the comparison of numerical results with data collected during the tests.

In order to investigate the effects of ventilation on Opalinus clay in the Mont Terri Underground Rock Laboratory (URL), a 10 m long section of an unlined microtunnel, 1.3 m in diameter, was sealed off in July 2002, and a ventilation system was installed in March 2003 (Mayor and Velasco, 2008). The microtunnel was raise-bored in February 1999. Between May and July 2002, the region 2 m in radius around the test section was instrumented with hygrometers, for measuring the relative humidity (RH), piezometers and extensometers. The RH is defined as the ratio of the partial pressure of water vapour in the mixture to the saturated vapour pressure of water at a given temperature. Several drilling campaigns were also run at different stages of the experiment to estimate the distribution of water content and chloride and sulphate concentrations around the microtunnel.

The VE was carried out in two phases. In the first phase, the VE tunnel was subjected to resaturation (11 months, starting from July 2002), then dry air was ventilated through the test section (8 months). In the second phase, a second resaturation and desaturation cycle was applied (11.5 and 20.5 months, respectively). The VE ended in September 2006. The significance of the study lies in the fact that all drifts and tunnels in the repository will be subjected to ventilation effects to some extent during the operative phase of the facility. In realistic operation scenarios, RH is estimated to be 60%–90% (Meier, 1998, Meier, 2004 and Gisi, 2007). During the drying cycles of the VE, the RH of incoming air varied between 2% and 30% to ensure stronger drying conditions over a limited period of time. It is believed that argillaceous rocks may be especially sensitive to the actions of this type. Specifically, the following issues are potentially involved:

1) Desaturation/resaturation of the rock, and phase changes;
2) Air/rock interface;
3) Damage/microcracking of the host rock due to hydro-mechanical and/or chemical effects;
4) Evolution of the excavation damaged zone (EDZ).

The DECOVALEX project (acronym for DEvelopment of COupled models and their VAlidation against EXperiments) was set up to support the development of computer codes and to compare model calculations with results from field and laboratory experiments. In the latest phase of the project (2008–2011), one of the tasks consisted of the analysis of the VE. In this paper, the general layout of the task and the main results are described:

1) The identification of relevant processes and of Opalinus clay parameters on the basis of the laboratory drying test.
2) Simple hydro-mechanical modelling of the VE up to the end of the first drying phase on the basis of the laboratory parameters and calibration of the models.
3) Advanced hydro-mechanical modelling of the VE, including blind prediction of the second drying period.

Specific issues linked to the task were treated in companion papers. One of the main difficulties experienced in the modelling exercises was the treatment of the evaporation boundary condition at rock–air interface. Bond et al. (2013a) reported the advances made by the different teams in that aspect. Millard et al. (2013) focused on the anisotropic nature of Opalinus clay and estimated the effects of anisotropic properties and stress state on the modelling results. The last companion paper (Bond et al., 2013b) described the chemical changes in argillaceous rocks resulting from ventilation and summarized the results obtained by the teams on conservative and reactive transport.

The participating modelling teams and the corresponding funding organizations are:

1) CAS (Chinese Academy of Sciences, China);
2) CEA (Commissariat à l'Energie Atomique) on behalf of IRSN (Institut de Radioprotection et de Sûreté Nucléaire, France);
3) Quintessa on behalf of NDA (Nuclear Decommissioning Authority) (UK);
4) UoE (University of Edinburgh) on behalf of NDA (Nuclear Decommissioning Authority) (UK);
5) JAEA (Japan Atomic Energy Agency, Japan).

MONT TERRI AND THE VENTILATION EXPERIMENT

The Mont Terri project started in 1995, as a collaboration between not less than eleven organizations from six different countries to investigate the hydrogeological, geochemical and geotechnical characteristics of an argillaceous rock formation (Opalinus clay). The project is based on the construction of a URL built as an extension of the security gallery of a highway tunnel between St Ursanne and Courgenay in Switzerland (Fig. 1). Four main construction phases were carried out: in 1996, several niches were excavated from the security gallery; in 1998–1999, a first gallery and the microtunnel in which the VE took place were excavated; and in 2004 and 2008, two new extensions were bored, respectively. Scientific experiments were set up from the different niches and galleries to investigate the rock response to excavation, ventilation, heating and bentonite emplacement (Thury and Bossart, 1999).

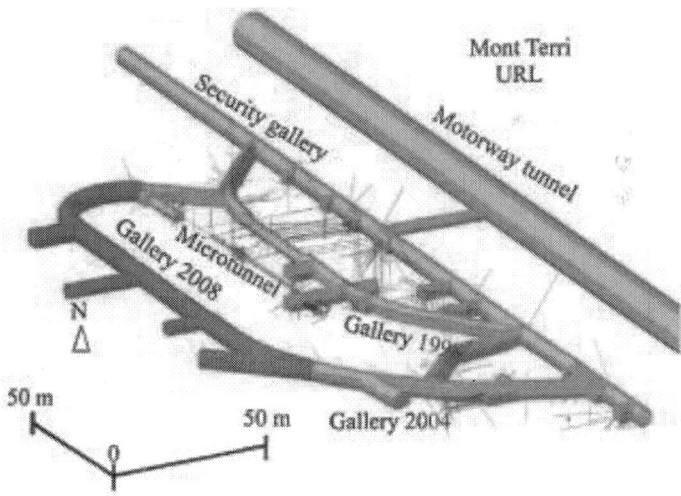

Figure 1. 3D view of the Mont Terri URL.

Opalinus clay

Opalinus clay is a stiff over-consolidated clay of Lower Aalenian age, corresponding to the Middle Jurassic. It was found in the Jura Mountains of Northern Switzerland. Its mineralogy consists mainly of sheet silicates (illite, illite-smectite mixed layers, chlorites, kaolinites), framework silicates (albites, K-feldspar), carbonates (calcite, dolomite, ankerite and siderite) and quartz (Bossart et al., 2002). There are three slightly different facies containing different mineral proportions: a shaly facies in the lower part of the deposit, a 15 m thick sandy-silty facies in the centre and a sandy facies interstratified with shaly facies in the upper part. The content of clay minerals may range from 40% to 80%, depending on the facies. The clay is sedimented in marine conditions, and its pore water is highly mineralized with total dissolved solids up to 20 g/L. This water contains a significant amount of seawater millions of years old. Total thickness is about 160 m.

In the location of the Mont Terri laboratory, the overburden depth varies between 250 m and 320 m (it is estimated that overburden reached at least 1000 m in the past). The laboratory is situated in an asymmetrical anticline formed during the folding of the Jura Mountains as shown in the geological cross-section of Fig. 2. The rock strata dip, because of the tectonic activity, with an angle of 30°–45° to the southeast. The bedding trace is perpendicular to the axis of the motorway tunnel. A picture of an Opalinus clay sample showing clearly the strong layering of the material is presented in Fig. 3.

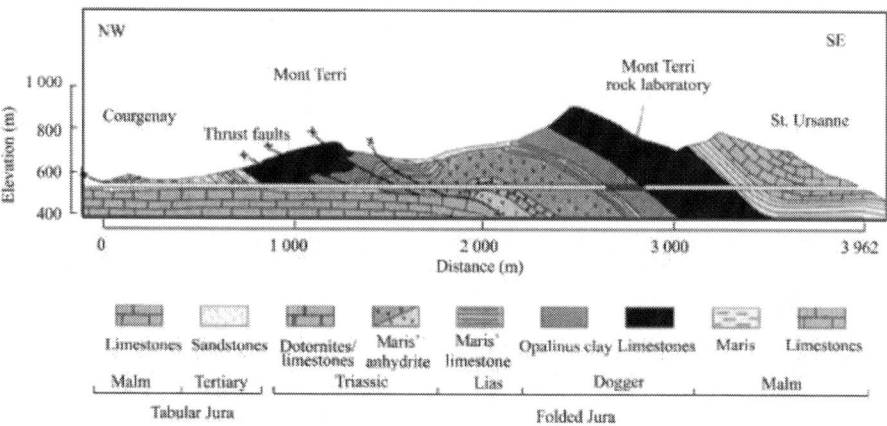

Figure 2. Geological profile in the vicinity of the Mont Terri underground laboratory as a function of the motorway tunnel metres (Schaeren and Norbert, 1989 and Thury and Bossart, 1999).

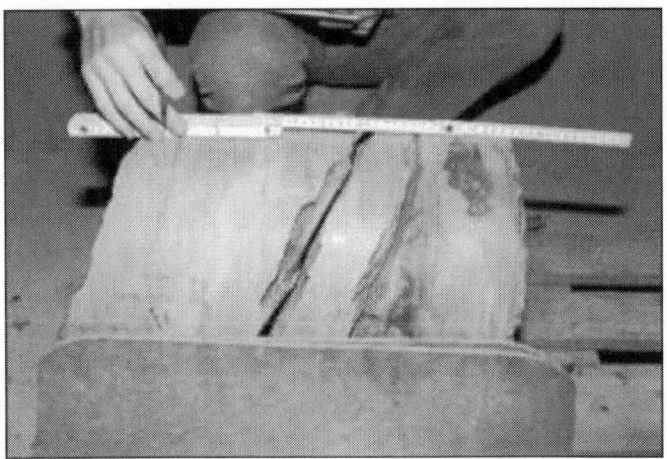

Figure 3. Opalinus clay sample.

Opalinus clay behaviour has been intensely studied by means of laboratory and in situ experimental programmes in the framework of the Mont Terri project. A general synthesis of the main physical and geotechnical parameters was reported in Bock (2001)and Wileveau (2005). Based on this information, the reference values of a series of parameters together with their likely ranges have been listed in Table 1. It can be noted that some of the parameters have different values depending on the orientation of the material, reflecting the anisotropy caused by the strong bedding of the clay. When information is scarce, no range estimation has been quoted. The following additional remarks can be made:

1) Reference elastic parameters have been based on measurements made in triaxial and uniaxial compression tests and on results of field dilatometer tests.
2) The reference shear strength parameters have been derived from laboratory triaxial tests. As strength depends on water content, only strength data for which the moisture content was within the range of the natural water content were considered.
3) Permeability measurements were made on laboratory specimens and using in situ boreholes. The dispersion of the in situ permeability measurements is illustrated inFig. 4, as this

parameter has an important influence on the results of the VE, which can be considered as a huge pump test itself.

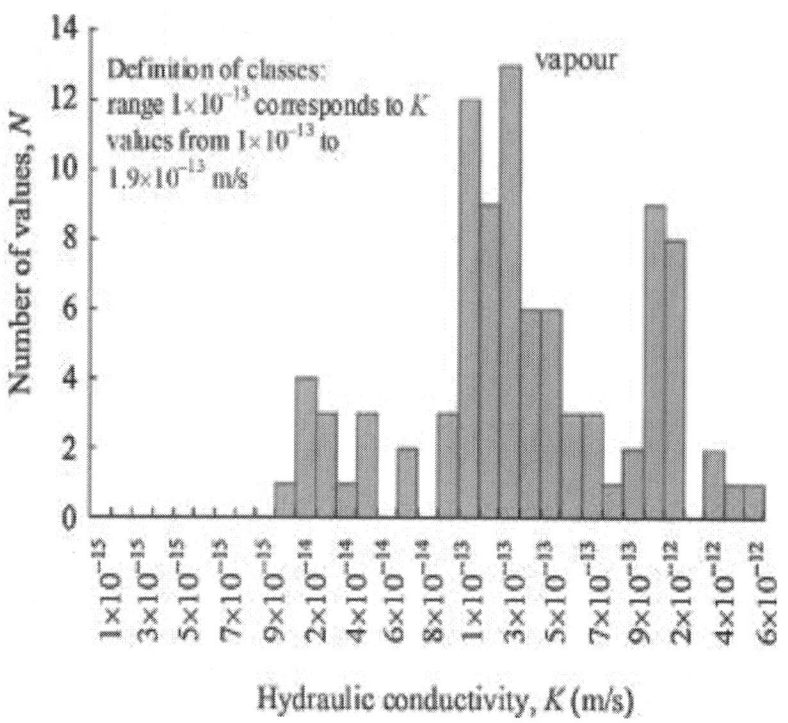

Figure 4. Compilation of results from field permeability tests performed on Opalinus clay in the Mont Terri underground laboratory (Nussbaum and Bossart, 2004).

4) Biot's coefficient is uncertain due to limited data.

Table 1. Reference parameters for Opalinus clay (Wileveau, 2005).

Mineralogy			Petrophysical properties			Hydraulic and hydromechanical properties	
Clay content (%)	Carbonate content (%)	Quartz content (%)	Density, ρ (g/cm³)	Water content, w (%)	Porosity, ϕ (%)	Water permeability of sound clay, k (m/s)	Biot's coefficient, b
62 (44–80)	14 (6–22)	18 (10–27)	2.45 (±0.03)	6.1 (±1.9)	15.7 (±2.2)	10^{-13} (10^{-12} to 10^{-14})	0.6 (0.42–0.78)

Mechanical properties											
Uniaxial compression strength (MPa)		Tensile strength (MPa)		Elastic modulus (MPa)		Poisson's ratio		Shear strength parameters			
$R_{c//}$	$R_{c\perp}$	$R_{t//}$	$R_{t\perp}$	$E_{//}$	$E_{\perp}$	$v_{\perp//}$	$v_{//}$	$c'_{//}$ (MPa)	$\varphi'_{//}$ (°)	$c'_{\perp}$ (MPa)	$\varphi'_{\perp}$ (°)
10 (±6)	16 (±6)	1	0.5	10,000 (±3700)	4000 (±1000)	0.24	0.33	2.2	25	5.0	25

Note: The symbols "//" and "⊥" refer to the material orientation parallel and perpendicular to the bedding, respectively. The data in brackets represent the range of the parameters.

The reference values have been used as a basis for estimating the parameters required in the numerical analyses reported below. However, alternative values have been adopted if information referring more specifically to the location of the ventilation in situ test was available.

A significant number of measurements using different procedures (borehole slotter, undercoring, and hydraulic fracturing) of the in situ stress have been made. They have been supplemented by geological observations and back analysis of instrumented excavations. A synthesis of the information available is reported in Wermeille and Bossart (1999) and in Martin and Lanyon (2003). The following comments can be made:

1) The major principal stress is subvertical and corresponds approximately with overburden weight (about 7 MPa).
2) The magnitude of the intermediate principal stress obtained from undercoring is consistent with the results from hydraulic fracture tests (about 5 MPa).
3) The value of the minor principal stress is quite low (about 2 MPa) and probably controlled by the presence of a deep valley to the SW of the laboratory. A low value of the minor principal stress is consistent with the small number of breakouts observed in vertical boreholes.

Technical features of the VE

The VE has been performed in a 10 m long section of the unlined raise-bored microtunnel (1.3 m in diameter), excavated in 1999 in the shaly facies section. The VE section is presented in Fig. 5. Different vertical planes (SAx, SBx, SCx and SDx) were instrumented along the test section.

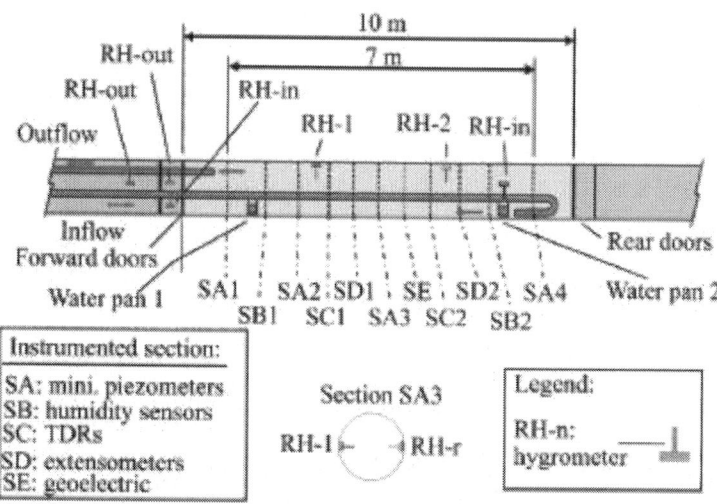

Figure 5. Layout of the VE.

The VE was sealed off by means of two double doors made of exotic wood (wengé) insensitive to RH variations (Fig. 6). The controlled ventilation during the test was accomplished using a system consisting of a blowing device that was located outside the test section (by a compressor, a drier and a bubbler), and inflow and outflow pipes that were equipped with flowmeters and hygrometers. Measurement of airflow mass (q_g) and RH of in- and out-going air allowed for establishing the global water mass balance of the test section according to

$$\theta_g^w = \omega_g^w \rho_g = \frac{RH}{100} \frac{p_{v0} M_w}{RT}$$

(1)

where θ_g^w is the volumetric mass of water in the gas phase, $\omega_g^w = m_w/m_g$ is the mass fraction of water in the gas phase, p_{v0} is the vapour pressure for saturated state, M_w is the molecular mass of water, R is the universal gas constant, T is the temperature, and ρ_g is the gas density (kg/m³). The water mass flux q_w (kg/h) is written as

$$q_w = \theta_g^w q_g$$

(2)

Figure 6. Picture of the test section.

The test history is illustrated in Fig. 7, in which the RH of the test section is plotted as a function of the time in the 12-year interval (149 months) between the excavation of the microtunnel and 2011. In this paper, we focus on the period between the excavation of the microtunnel and the end of the controlled ventilation. Several phases can be usefully distinguished:

1) Phase 0, in which the VE tunnel was excavated and left open without controlled ventilation conditions (from February 1999 to July 2002). This phase lasted for about 3.5 years (41 months).
2) Phase 1, in which the VE tunnel was subjected to controlled ventilation conditions resulting in rock resaturation (from July 8th 2002 to May 28th 2003) and subsequent desaturation (from May 28th 2003 to January 29th 2004). This phase lasted for about 1.5 years (19 months).
3) Phase 2, in which an additional episode of controlled resaturation (from January 29th 2004 to July 11th 2005) and

desaturation (from July 11th 2005 to September 24th 2006) was performed. This phase lasted for about 2.5 years (33 months).

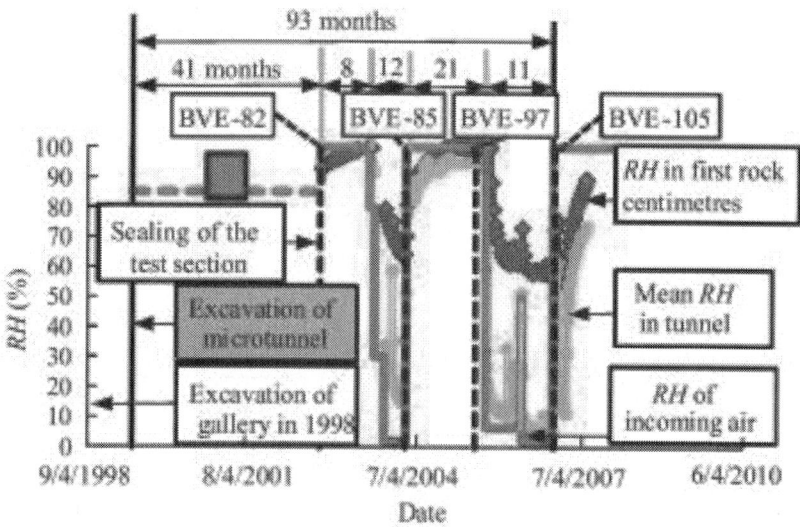

Figure 7. Relative humidity history of the test section.

Finally, a last resaturation stage was applied.

In Phase 0, representative of natural ventilation of a niche (tunnel is closed at one end), the RH in the tunnel was not measured. Nevertheless, some relative humidity measurements are available at several locations in the URL. Meier (1998) measured the RH in gallery and came up with values around 50%. Meier (2004) and Gisi (2007)measured the RH in boreholes. They monitored values of 80%–90% and 75%–95%, respectively. As the microtunnel is closed at one end, its configuration is similar to that of a borehole. Moreover, Garitte and Gens (2011) back calculated the value of the RH in the tunnel (85%) in this period to reproduce correctly the pore water pressure and the water content gradient around the tunnel before the controlled ventilation period.

Phase 1 corresponds with the VE itself and starts with the sealing of the test section with two double doors. Only by isolating the section, the measured RH in the tunnel approaches 100%. The RH in the first rock centimetres increases progressively. After 11 months, the rock mass is believed to be saturated and the first drying phase starts. RH of blown air is as low as 2% and that in the tunnel decreases down to 15%. In the first rock centimetres, RH decreases to 65%. The resaturation–desaturation process is then repeated in Phase 2.

Several drilling campaigns were carried out at key moments to obtain radial profiles of water content and chloride and sulphate concentrations:

1) Borehole BVE-82 on July 5, 2002 before isolation of the test section.
2) Boreholes BVE-85 and BVE-86 on January 26, 2004 at the end of the first desaturation.
3) Boreholes BVE-97, BVE-99 and BVE-100 on May 1, 2005 at the end of second resaturation.
4) Boreholes BVE-105, BVE-106, BVE-107, BVE-109 and BVE-110 on October 4, 2006 at the end of the second desaturation.

The time of the drilling campaign is indicated in Fig. 7 by vertical dotted lines. The information from borehole BVE-82 and from the piezometers installed at that time was used to back calculate the most likely value of RH in the microtunnel between its excavation and the sealing off of the test section. The instrumentation system installed in the VE area is summarized in Table 2.

Table 2. Summary of the emplaced sensors.

Measurement item	Sensors (boreholes) number	Min. and max. distance to tunnel wall (m)
RH measurement in the tunnel	10	Up to 0.02
Water evaporation in water pans	2	–
RH measurements in the rock mass	24	0.25 (0.07)–1.5
Water balance of the system	–	–
Water content profiles	13	0–1 (1.5)
Chloride and sulphate concentration profile		
Water pressure evolution	28	0.5–1.5 (5.6)
Relative displacements	8	Between wall and 2 m inside

Note: Before Phase 2, new instrumentation was emplaced extending the instrumented range. The new range is indicated by the values between brackets.

Different geophysical methods (seismic refraction, interval velocity and cross-hole measurements) were applied to evaluate the extent and the development of the EDZ around the microtunnel (Schuster, 2007). The extent of the EDZ was estimated to be 5–25 cm in the bedding planes direction and 10 cm in the perpendicular direction. No significant changes induced by ventilation were identified and the EDZ was thus related to stress redistribution during the excavation exclusively. Moreover, observation of the tunnel wall showed that the state of the rock surface in the test section before and after the two drying phases was fairly good: neither relevant rock failures nor far-reaching cracking were observed on the walls (Mayor and Velasco, 2008). The small displacements registered in the extensometers during the different cycles tend to confirm the fact that the EDZ does not develop during drying and wetting. Garitte and Gens (2011) predicted an extent of the EDZ (by modelling) of about 30 cm. According to this modelling, the EDZ was intensified during the ventilation cycles but not extended significantly.

The laboratory drying test (Floría et al., 2002) was performed in 2001, before the start of the in situ VE. Three cylindrical Opalinus clay samples (0.28 m in height and 0.1 m in diameter) and a water pan were placed on balances in a drying chamber. The layout of the laboratory drying test is presented in Fig. 8.

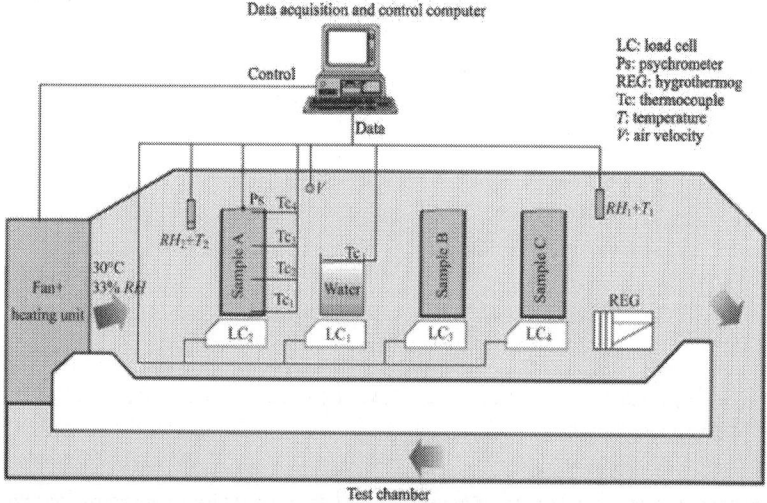

Figure 8. Layout of the laboratory drying test.

All sample walls, except the top, were isolated in order to create conditions as close as possible to one-dimensional (1D) flow. The atmospheric conditions applied through a fan and heating unit were measured in various sensors inside the chamber. The values of RH (measured on top of the samples) varied between 20% and 50%. A constant temperature of 30 °C was maintained. The air velocity applied was between 30 cm/s and 70 cm/s. The evaporation rate and water loss from a water pan were also measured to act as a reference.

Continuous weight measurement allowed a direct estimation of the water loss in each of the samples. A perhaps fortuitous correlation should be noted: the sample closest to the fan heating unit loses most water and the farthest one the least. After 142 days of testing, sample A had lost about 150 g from the initial 352 g. Complementary information about the water content distribution was obtained at 21 and 99 days by withdrawing a sample at each of those times, cutting it into 3 cm slices and measuring the water content of each slice. Integrating the water content profiles, we found a water loss of 59 g, 121 g and 151 g for samples C, B and A, respectively. That corresponds remarkably well with the water loss obtained through the continuous weight measurements, which increases the confidence in the experiment.

THEORETICAL FORMULATIONS AND PARTICIPANTS

Only a summary of the different formulations used by the teams is reported herein. The different equations resolved by the teams and their corresponding codes are given inTable 3. Most of the codes, FRT-THM, CAST3M (Verpeaux et al., 1989), THAMES (Chijimatsu et al., 1998) and RockFlow/Geosys are finite element codes and only QPAC (Maul, 2010), the code used by Quintessa, is a finite volume code. There was a common agreement that the water mass balance was the most important equation by far. Air mass and energy balances were found to have a negligible influence as the gas pressure and the temperature were constant throughout the test. Deformation of the porous medium (stress equilibrium) affects the water mass balance, but its effects may be categorized as second order (Millard et al., 2013). Solute transport and reactive transport were affected by the desaturation of the rock mass around the microtunnel, but the back coupling was estimated to be negligible (Bond et al., 2013b).

Table 3. Summary of the codes and balance equations used by the modelling teams.

Team	Code	Water balance	Solid balance	Air mass balance	Energy balance	Stress equilibrium	Solute transport	Reactive transport
CAS	FRT-THM	X	X	X	X	X	X	
CEA	CAST3M	X	X			X	X	
JAEA	THAMES	X	X		X	X	X	
Quint	QPAC	X	X	X	X	X	X	X
UoE	RocFlow/ Geosys	X	X	X			X	

Common basis

The previous considerations lead us to focus on the equation of the water mass balance:

$$\frac{\partial}{\partial t}(\theta_1^w S_1 \phi + \theta_g^w S_g \phi) + \nabla \cdot (j_1^w + j_g^w) = f^w$$

(3)

where ϕ is the porosity, θ_1^w is the volumetric masses of water in the liquid phase; S_1 and S_g are the degrees of saturation of liquid and gas phases, respectively; j_1^w and j_g^w are the mass fluxes of water in the liquid and gas phases, respectively; $\theta_1^w = \omega_1^w \rho_1$, where $\omega_1^w = m_w/m_1$ is the mass fraction of water in the liquid, and ρ_1 is the density of liquid.

We distinguish the changes of water mass in a certain volume due to changes in porosity, degree of saturation, and liquid and gas densities with respect to time in the first term and the divergence of water fluxes in the second term. The third term is a sink/source term which is equal to 0 in the present problem.

The advective flow of water in the liquid phase j_1^w (kg/m² s) is

$$j_1^w = \theta_1^w q_1 \tag{4}$$

where q_1 is the Darcy velocity (m/s), i.e. the volumetric flow/section, which is proportional to the water pressure gradient ∇P_1 (Pa/m):

$$q_1 = -\frac{k k_{rl}}{\mu_1} \nabla P_1 \tag{5}$$

where k is the intrinsic permeability (m²), μ_1 is the dynamic viscosity (Pa s), and k_{rl} is a coefficient depending on the degree of saturation.

The dependence of the permeability on the degree of saturation is introduced through

$$k_{rl} = \sqrt{S_l}[1 - (1 - S_l^{1/\lambda'})^{\lambda'}]^2 \tag{6}$$

known as Van Genuchten law, where λ' is a shape parameter. The transport of water in the gas phase can be decomposed into

$$j_g^w = (i_g^w)_{\text{advection}} + (i_g^w)_{\text{diffusion}} + (i_g^w)_{\text{dispersion}} \tag{7}$$

where the first term in the right represents the flux of water by motion of the gas phase and the second term in the right the flux of water by diffusion of water vapour inside the gas phase (non-advective flow). Dispersion was neglected. Gas motion was found to be negligible as gas pressure is constant throughout the test. Vapour diffusion is expressed by Fick's law:

$$(i_g^w)_{\text{diffusion}} = -(\phi \rho_g S_g D_g^w I) \nabla \omega_g^w \tag{8}$$

where D_g^w is the vapour diffusion coefficient (m²/s); and $\nabla \omega_g^w$ is the gradient of vapour concentration, defined as the ratio of the mass of water in the air to the mass of air. Vapour diffusion was found to have a significant influence on the results. The relationship between suction $(P_g - P_1)$ and the liquid degree of saturation is idealized by the modified Van Genuchten retention curve:

$$S_e = \frac{S_l - S_{rl}}{S_{ls} - S_{rl}} = \left[1 + \left(\frac{P_g - P_l}{P}\right)^{(1/(1-\lambda))}\right]^{-\lambda}\left(1 - \frac{P_g - P_l}{P_s}\right)^{\lambda_s} \tag{9}$$

where λ is a shape parameter, S_{rl} is the residual saturation, S_{ls} is the maximum saturation; P_s and λ_s are two material parameters, and P corresponds with the air entry value.

For suction values lower than the air entry value, water is retained in the clay pores although the pore water is submitted to tension. When suction exceeds the air entry value, the porous material starts to desaturate. This allows the distinction between a suction limit and a desaturation limit. UoE used another relationship (Ippisch et al., 2006). Finally, we make the hypothesis that water in the pores, in vapour form and in liquid form, is in equilibrium, through Kelvin's law that relates water content in the gas phase to the suction in the liquid phase:

$$\theta_g^w = \omega_g^w \rho_g = (\theta_l^w)^0 \exp\left[\frac{-(P_g - P_l)M_w}{(273.15 + T)R\rho_l}\right] \tag{10}$$

The relative humidity can be related to θ_g^w through

$$RH = \frac{p_v}{(p_v)_0} \times 100\% = \frac{\theta_g^w}{(\theta_g^w)_0} \times 100\% \tag{11}$$

where p_v is the vapour pressure, and subscript $()_0$ stands for saturated state.

Conceptual models and specific features
A considerable amount of computation was run by the different teams for the laboratory drying test and for the VE. In both cases, 1D, 2D (two-dimensional) and 3D (three-dimensional) models were developed.

For the laboratory drying test, the first modelling iteration round showed that the 1D hypothesis was appropriate enough to tackle the case. The samples were considered as being initially saturated (7.14% of water content) and in equilibrium with the atmospheric pressure (pore water pressure and total stresses of 100 kPa). For the in situ VE, 1D modelling was shown to provide satisfactory results. In this case, the tunnel wall was considered as a boundary condition. Obviously, treatment of the anisotropy required an upgrade of the model to 2D (plane strain). Improved treatment of the evaporation boundary condition, in which the in- and out-pipes were considered as model boundary and the air flow in the tunnel was modelled (Bond et al., 2013a), implied the development of 3D models. The initial conditions considered a water pressure of 1.85 and 2.0 MPa and a stress state of 4.5 MPa, both representative of the situation at Mont Terri (Gens et al., 2007). The excavation and history of the microtunnel were simulated.

VEs are intended to investigate how an initially saturated porous medium dries when submitted to a dry environment. A key issue related to that is the exchange of water between the rock and the dry environment. In this kind of low porosity material, water is believed to leave the rock as vapour. Evaporation, the process by which molecules in a liquid state spontaneously become gaseous, occurs at rock–air interface and in the desaturated porous material near the evaporation boundary. Hereby the relative humidity profile, a measurement of the amount of water vapour that exists in a gaseous mixture of air and water, may be continuous along the rock–air interface or discontinuous as a consequence of the nature of the interface, the wind velocity in the tunnel and the water availability in the rock (that depends on the pore size, pore distribution and the degree of saturation of the interface pores). Although the evaporation boundary condition has been treated in detail in Bond et al. (2013a), a summary is given here.

The water loss from the pans and the samples in the laboratory drying experiment is used to discuss the evaporation boundary condition (Fig. 9). The water loss from the pans acts as a reference for free surface evaporation. At 20, 100 and 140 days, we know the measured degree of saturation on top of the samples as at each of these times one sample has been taken out of the chamber. The equivalent free water surface on top of the sample may be estimated by multiplying the top area with the porosity and the degree of saturation. Multiplying the water loss from the pans by that factor, we obtain an equivalent water loss that may be compared to the water loss from the samples (triangular symbols in Fig. 9). This comparison seems to indicate that (at least) in the first 20 days of the experiment, the water on top of the sample behaves as free surface water. Later on a clear deviation from that behaviour is observed. This deviation reflects the imperfectness of the air–rock interface.

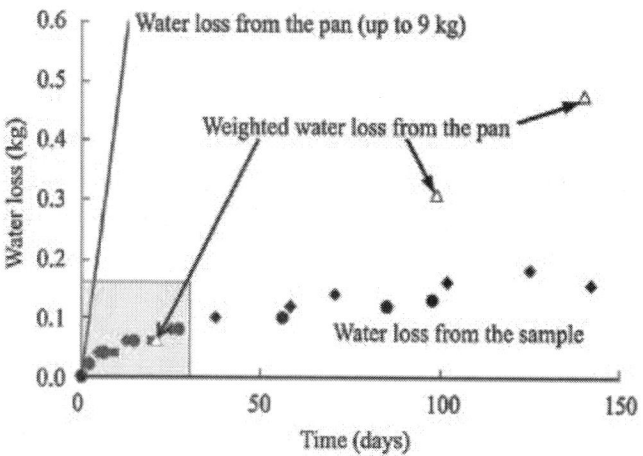

Figure 9. Measured water loss in the laboratory drying test: from the water pan and the samples.

The numerical treatment of the evaporation boundary condition was done using three different approaches. The most straightforward is to apply directly the measured relative humidity (on top of the sample or on the tunnel wall):

$$RH_{boundary} = RH_{measured} \tag{12}$$

According to the water loss measurements in the laboratory experiment (Fig. 9), this option is only adequate for the first days of the experiment when the sample is almost saturated. Nevertheless, different computations have shown that using this option for the entire duration of the laboratory experiment does not induce a significant error. For the in situ ventilation instead, applying directly the relative humidity was shown to provide unsatisfactory results. The teams using this option had to apply a RH value between the RH measured in the tunnel and that measured in the first rock centimetres. JAEA, for instance, determined an empirical relationship between RH in the gallery and RH to be applied.

The second option consists in applying a water vapour flux as

$$j_g^w = \beta_g[(\rho_g \omega_g^w)^0 - (\rho_g \omega_g^w)] \tag{13}$$

where the superscript $(\cdot)^0$ stands for the prescribed values, and β_g is a penalty or interface coefficient (m/s) that controls the velocity at which the contour values tend to the prescribed values. For low β_g values, equilibrium is not reached instantaneously and less water is extracted. For high β_g values, Eq. (13) degrades to Eq. (12). The penalty coefficient may depend on degree of saturation and/or porosity of the interface elements. This allows for having high values when the rock surface is saturated and low values when water availability is low. Constant β_g values were shown to be sufficient to reproduce correctly RH in the first rock centimetres applying RH measured in the tunnel.

It was shown that using Eq. (13), it was possible to tackle the interface problem in a satisfactory way, but the observations of RH in the tunnel during the VE had to be used. Quintessa and CAS (Bond et al., 2013a) succeeded in developing an innovative approach (Option 3), consisting in modelling the tunnel air flow and the rock mass, and the boundary condition being displaced from the tunnel wall to the in-pipe. Prescribing the incoming air flow and its RH, they succeeded to predict the value of RH in the tunnel, in the first rock centimetres, the rock behaviour and RH of the out-coming air at the out pipe.

In this section, we outlined the difficulties of the evaporation boundary condition and solutions employed to treat the problem. Evaporation does not occur only at the rock–air interface, but also in the desaturated ring around the microtunnel. In that zone, water transport occurs by (1) advection (Darcy's law) and (2) non-advective vapour diffusion. Vapour diffusion was only considered by CAS, CEA and Quintessa. JAEA, UoE and an additional computation by CEA(2) neglected vapour diffusion. The difference between the two approaches resulted in consistent differences in modelling results and parameters determination. Note that the vapour concentration gradient drives water vapour towards the air–rock interface and contributes to the water vapour amount available for leaving the rock.

MODELLING RESULTS

Laboratory drying test

The modelling work started with the laboratory drying test because of its similarity with the VE in situ test and of its simple configuration. This test allowed for evaluating the capabilities of the different codes and the relative importance of the processes in place. In the first modelling iteration, most of the input parameters were taken from Table 1, although some of them had to be updated according to data specific to the test area. In the second modelling iteration, some of the parameters were calibrated. Sensitivity analyses have been presented in Garitte et al., 2009 and Garitte et al., 2010. These simulations showed the importance of: (1) the value of water permeability and its dependency on saturation, and (2) water transport through vapour diffusion. Point 1 was used to calibrate the model. The first step of the calibration consists in fitting the first days of the experiment to obtain the saturated permeability. The remaining time of the experiment is then used to calibrate the dependency of permeability on degree of saturation.

Physical properties (porosity, density, water content, etc.) were taken directly from reports describing the results of the VE drilling campaigns (Traber, 2003, Traber, 2004 and Fernández et al., 2006). Measurements of water retention curves from four independent laboratories were used (Gens, 2000, Muñoz et al., 2003, Zhang and Rothfuchs, 2005 and Villar, 2007) as an input for the models (Fig. 10). The retention curve was shown to have a significant influence on the results.

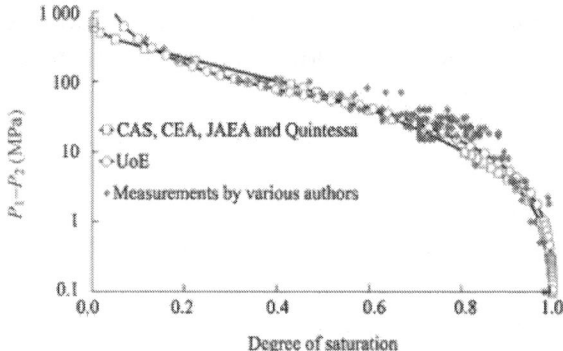

Figure 10. Retention curves of Opalinus clay: measurements by various authors (Gens, 2000, Muñoz et al., 2003,Zhang and Rothfuchs, 2005 and Villar, 2007) compared with the functions used by the different modellers.

Comparison of the modelling results with the laboratory measurements has been organized in two groups: (1) the teams considering water transport through advective flux of liquid water and diffusion of water vapour (CAS, CEA(1) and Quintessa), and (2) the teams considering only advective flux as water transport mode (CEA(2), JAEA and UoE). Comparison of the water loss throughout the experiment is presented in Fig. 11. Simulation results produced by five different codes, based on independent parameters determination, lie within the measured variation range which is an achievement in itself.

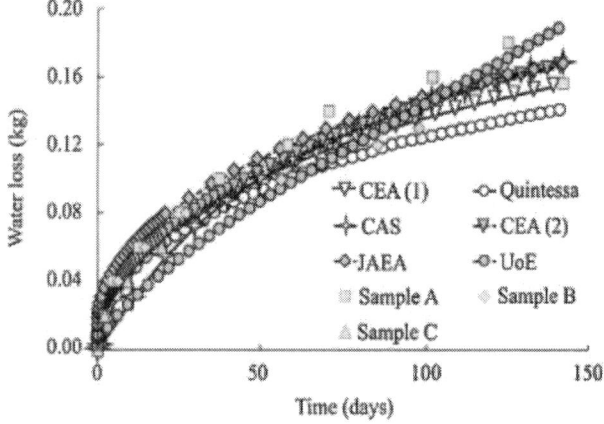

Figure 11. Measurement and simulation of the water loss history.

The water content profiles at 21, 99 and 142 days are presented in Fig. 12 and Fig. 13, for CEA(2), JAEA and UoE and for CAS, CEA(1) and Quintessa, respectively. Both groups achieved not only quite a good agreement with the measurements, but also between themselves. For instance, the parameters sets used by CEA(2) and JAEA are very similar and this is reflected in the similarity of their modelling results. Further, we observe a good consistency between the continuous water loss results and water content profiles. For example, Quintessa's water loss results are in perfect agreement with the measurements from sample B that was retrieved from the chamber after 99 days, but overestimates sample C and underestimates sample A. The same tendency is observed in the water content profiles.

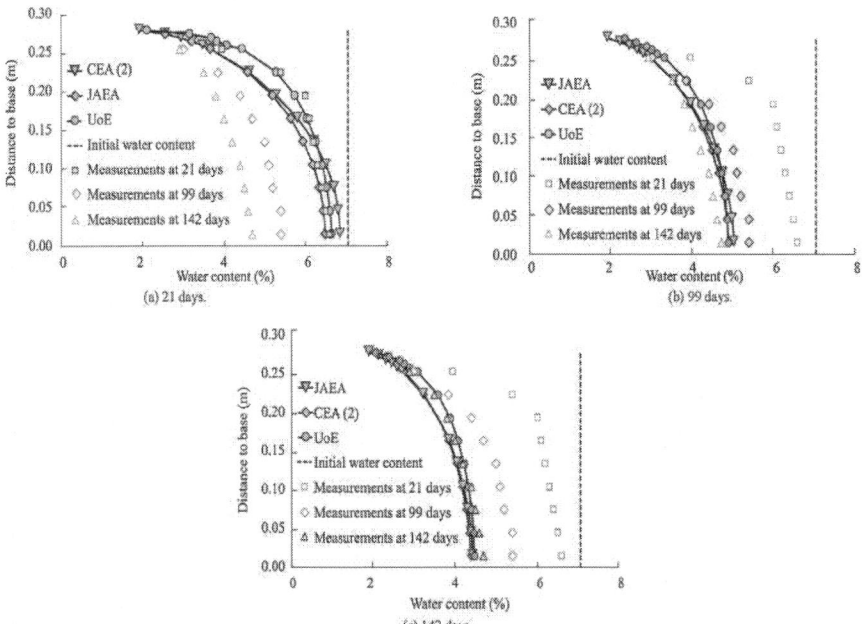

Figure 12. Comparison of measured and simulated water content profiles in the sample taken out at various days for the teams that did not consider vapour transport.

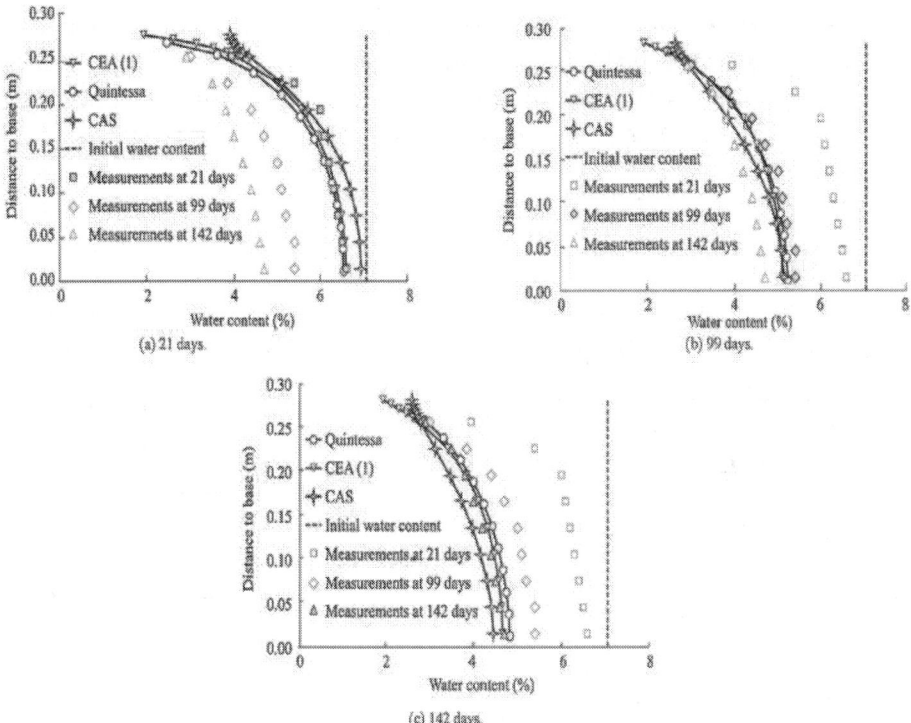

Figure 13. Comparison of measured and simulated water content profiles in the sample taken out at various days for the teams that did consider vapour transport.

The dependency of water permeability on degree of saturation used by the different teams is illustrated in Fig. 14. Clearly, we observe two tendencies. CAS, CEA(1) and Quintessa (the teams considering vapour diffusion) had to use a relatively high saturated permeability value that decreases fast with decreasing saturation. That is because in the desaturated zone, part of the water transport occurs by vapour diffusion.

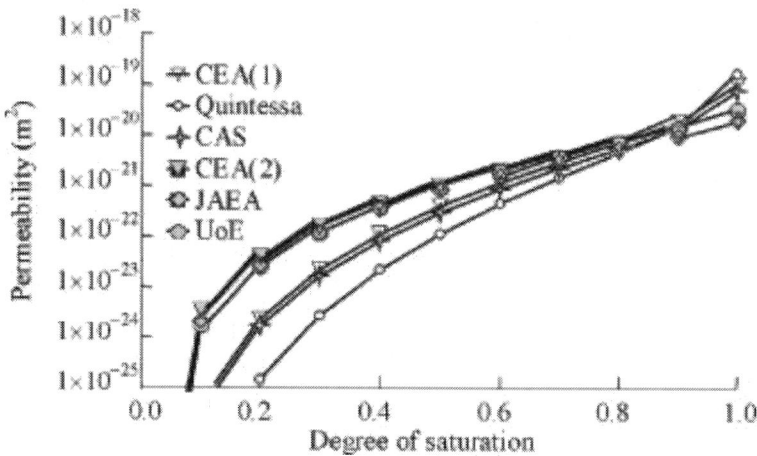

Figure 14. Water permeability in function of the degree of saturation as used by the different teams.

The relative importance of water diffusion and advective transport is illustrated in Fig. 15, representing simulated profiles of vapour to liquid flux ratio at different times. In the first days of the experiment, water transport occurs only through advection of liquid water. At this stage, water permeability used by the groups considering vapour diffusion is high and allows reproducing well the initially fast water loss. However, water transport quickly becomes dominated by vapour diffusion in the top centimetres of the sample as this zone desaturates. At the end of the test, the desaturation front reaches the bottom of the sample where the ratio grows up to 50%. The parameter sets determined by the different teams are summarized in Table 4.

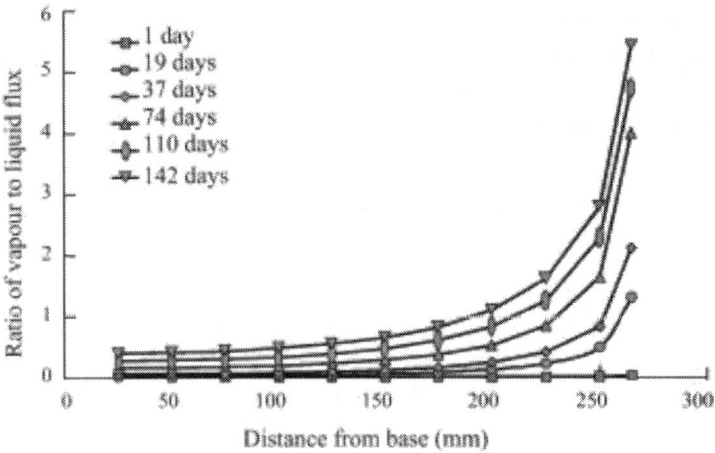

Figure 15. Profiles of vapour to liquid flux ratio at different times (calculated by Quintessa).

Table 4. Summary of the relevant parameters used by the different teams.

Team	Solid grain density, ρ_s (kg/m³)	Porosity, ϕ	Intrinsic permeability, k (m²)	Dynamic viscosity, μ (Pa s)	Liquid relative permeability, λ'	Vapour diffusion coefficient, D_g^w (m²/s)
CAS	2710	0.165	7.5×10^{-20}	1×10^{-3}	0.4	6×10^{-6}
CEA(1)	2710	0.16	1×10^{-19}	8×10^{-4}	0.4	2.6×10^{-5}
CEA(2)	2710	0.16	2×10^{-20}	8×10^{-4}	0.68	
JAEA	2743	0.162	2×10^{-20}		0.65	
Quint.	2700	0.16	1.7×10^{-19}		0.3	5×10^{-6}
UoE		0.19	3.2×10^{-20}	1×10^{-3}	0.6	

Team	Young modulus, E (GPa)	Poisson's ratio, ν	Air entry value (RC), P_0 (MPa)	Shape parameter (RC), λ	Maximum suction (RC), P_s (MPa)	2nd shape parameter (RC), λ_s
CAS	6	0.27	3.9	0.128	700	2.73
CEA(1)	6	0.27	3.9	0.128	700	2.73
CEA(2)	6	0.27	3.9	0.128	700	2.73
JAEA			8	0.15	700	2.73
Quint.	2.5	0.3	3.9	0.128	700	2.73
UoE			3.9	0.44	[a]	3.6×10^{-8} [a]

[a]UoE used a different equation for the water retention curve (Brooks and Corey, 1966).

In Fig. 16, the hydraulic conductivity values determined by the different teams on the basis of the laboratory drying test are compared to the geometric mean and the range of all the in situ measurements made in Mont Terri (Fig. 4) and to the pulse test measurements made in the piezometers of the VE (Mayor and Velasco, 2008). It is remarkable that the teams considering vapour diffusion have come up with values very close to the ones measured in the VE site. Hydraulic conductivity measurements in the VE test site are situated in the higher range in Mont Terri. Note that there is no clear indication of increasing permeability towards the tunnel, which may have been the case if an important EDZ is presented. A constant saturated water permeability was thus adopted as a starting point for the modelling of the VE.

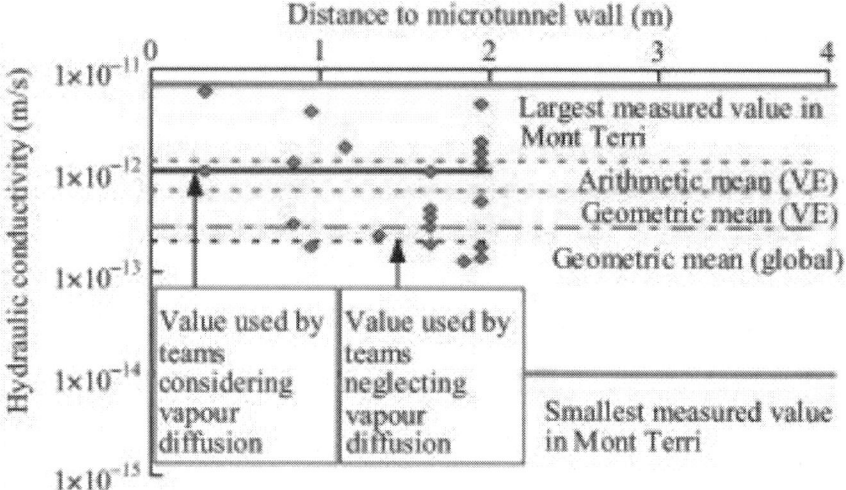

Figure 16. Hydraulic conductivity values measured in the piezometers of the VE before the experiment (Mayor and Velasco, 2008). The geometric mean from all in situ measurements (Nussbaum and Bossart, 2004) was added.

VE modelling on the basis of laboratory determined parameters
The information collected in the analysis of the laboratory drying test was injected in the modelling of the in situ VE to assess the predictive capacities. At this stage, the teams were asked to compare their modelling results using the parameters determined in the laboratory drying test with the measurements from the first ventilation phase. The modelling results from the different teams may be considered as a benchmark between the teams as well as a validation exercise of the models against the measurements. After this first step, the models were calibrated in function of the

measurements from Phase 1 and a sensitivity analysis was run to evaluate the current formulations capacities and seek for possible explanation for discrepancies between models and measurements. New elements were introduced and the acquired expertise was used to make a blind prediction of Phase 2, as described in the next section.

Pore water pressure, water content and relative humidity measured just before sealing the test section are plotted as a function of the distance to the tunnel axis in Fig. 17. The red straight line corresponds to the atmospheric pressure. Below this value, the pore water is likely to be in suction and the piezometers will cavitate (not lower pressure will be measured). The modelling results from the teams are added. According to the water content measurements, the partial desaturation limit lies between 13 and 31 cm from the wall. Behind 31 cm, the variation of water content is probably associated with measurements scatter. This trend is confirmed by the RH measurements that show a scatter between 95% and 100% behind 31 cm. 95% was estimated to be the in other galleries. The computation results from the teams provide quite a good agreement with the measurements considering the predictive nature of the exercise. The relatively good distinction between the partial desaturation limit and the suction limit reproduced by the models should be highlighted, although that achievement is strongly related to the care yielded to the calibration of the retention curve (Fig. 10).

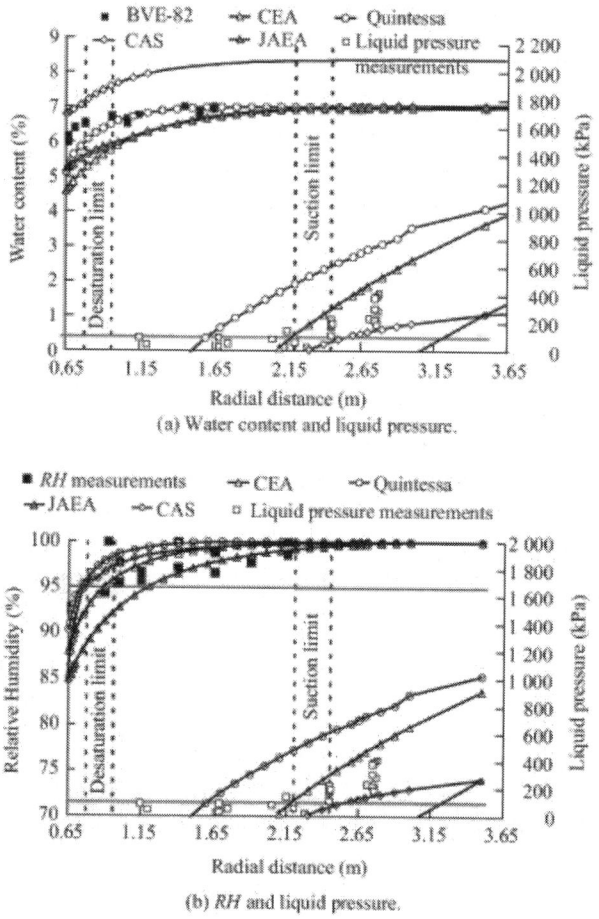

(a) Water content and liquid pressure.

(b) RH and liquid pressure.

Figure 17. Measured and simulated profiles of water content (BE-82) for RH and pore water pressure just before the sealing of the test section (5 July 2002).

Measurement of the global water mass balance of the in situ experiment is probably one of the main achievements of the VE as back calculation allows us to consider the experiment as a pump test involving an important volume of rock over a large duration. Indeed, in situ measurement of water permeability in low permeability porous medium is still a challenging issue as the technical problems encountered in conventional small scale-short duration pulse (or shut in) tests are not entirely resolved: injection of very small quantities of water, contact between the porous material and the injection chamber, and the volume contrast between the injection chamber itself and the pore volume. In the case of the VE, back calculation of the water permeability is obviously influenced by the estimation of the relative permeability and the water vapour diffusion in the partially desaturated zone and by advective water transport in the zone in suction where the validity of Darcy's law may be questioned.

The measured water mass balance and the modelling results are compared in Fig. 18. In this figure and in most other time evolution figures in this paper, the left hand scale represents the variable we are studying and the right hand scale represents the RH of incoming air to act as a time reference. Most of the teams reached a good agreement with the measurements. The prediction made by CEA should be highlighted: the agreement with the measurements was perfect. Their modelling results covered the measurement points. Direct application of the RH on the tunnel wall (Option 1 in Section1), used by CAS and JAEA, seemed to produce an abrupt response to ventilation regime changes. This effect appeared to be attenuated by the use of a flux condition and a penalty coefficient (by CEA and Quintessa).

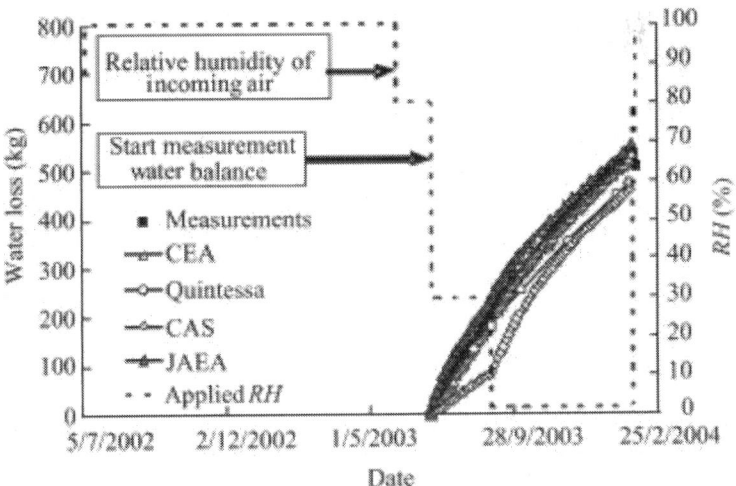

Figure 18. Measured and simulated water mass balance of the test section.

The RH measured in the tunnel is presented in Fig. 19(a). The RH calculated by the teams on the tunnel wall is added. For JAEA this coincides with the applied RH which is the RH of the incoming air (Option 1). CAS applied a RH somewhat higher than that of the incoming air RH and used also Option 1. CEA and Quintessa applied the RH of incoming air (the same as JAEA), but used a flux boundary condition (Option 2) and predicted a RH on the tunnel wall somewhat higher than mean RH measured in the tunnel in agreement with the discontinuous RH profile along the air–rock interface. The approach used by CEA and Quintessa allowed them to predict the evolution of RH in the first rock centimetres (Fig. 19(b)). CAS and JAEA instead overestimated the decrease of RH in the first rock centimetres and the response to ventilation regime change was too pronounced, which is probably due to the direct application of the RH on the tunnel wall (Option 1).

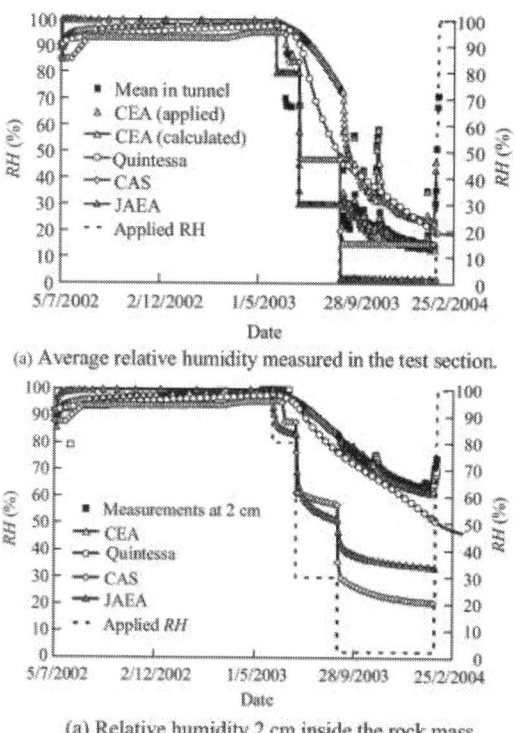

(a) Average relative humidity measured in the test section.

(a) Relative humidity 2 cm inside the rock mass.

Figure 19. Relative humidity in the tunnel and in the first rock centimetres.

At 25 cm (Fig. 20(a)), the tendency of the measurements is also better reproduced by the teams considering the flux condition. The situation is reversed at 35 cm (Fig. 20(b)), although it should be noted that the measured values are very close to 100% and hence not totally reliable.

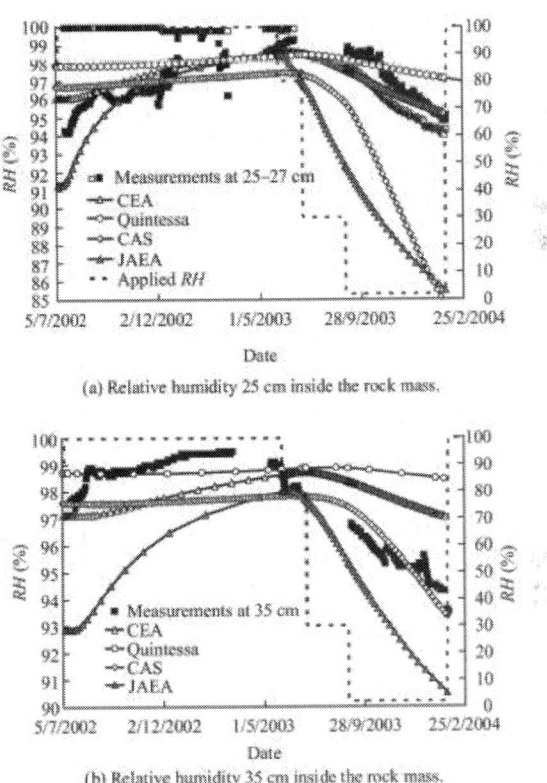

(a) Relative humidity 25 cm inside the rock mass.

(b) Relative humidity 35 cm inside the rock mass.

Figure 20. Measured and simulated relative humidity of various points inside the rock mass.

The measurements from the 24 piezometers are plotted in Fig. 21. Obviously, most of the piezometers do not monitor any response at all as before the start of the VE, all sensors closer to 1.8 m from the tunnel wall are situated in the suction zone. Those between 1.8 and 2.1 m from the wall exhibit an initial pore water pressure between 120 and 800 kPa. Once the drying phase starts, they respond quickly. Although the simulations predict quite well the pore water pressure state before the drying phase, they do not succeed in reproducing the rapid pore water pressure drop when drying starts. This consistent failure to reproduce this phenomenon by all the teams suggests that some process in the modelling is missing. Given the rapidity of the pressure drop, we are inclined to believe that it is caused by some mechanical response of the rock to drying as hydraulic dissipation would involve a much longer response time.

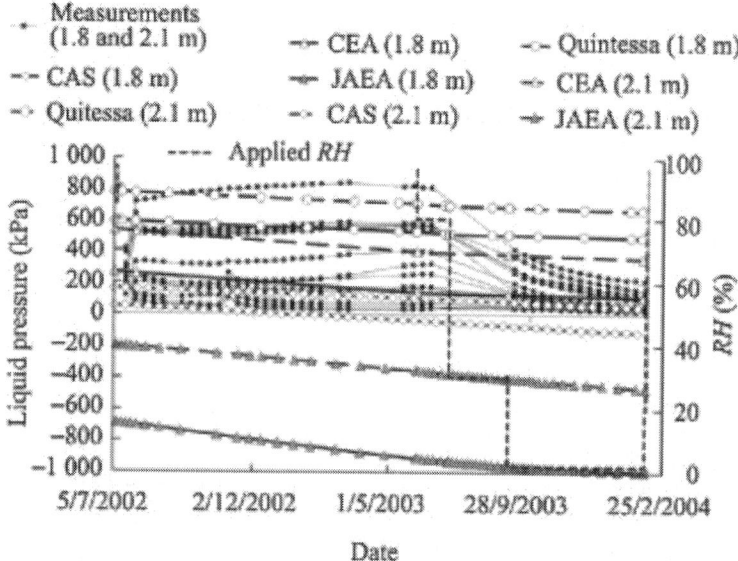

Figure 21. Measured and simulated pore water pressure between 1.8 and 2.1 m from the microtunnel wall.

Eight extensometers were installed in the test section. They show that wetting induces an expansion of the rock mass and drying a compression (Fig. 22). The simulations represent quite well the tendency that is explained by the changes in suction during the drying–wetting cycles that affect the effective stress. The strain changes induced by the wetting-drying cycles have a small amplitude compared to other processes like excavation.

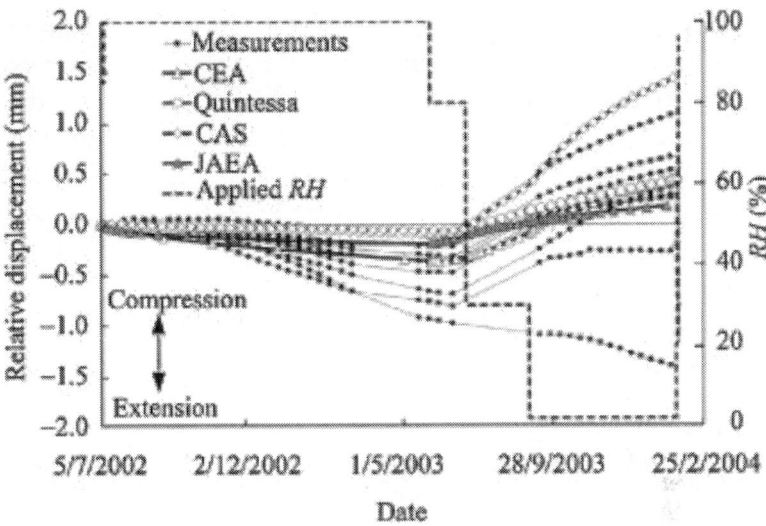

Figure 22. Measured and simulated relative displacement between the microtunnel wall and points 2 m inside the rock mass.

The final state of the rock mass at the end of the first ventilation phase is summarized inFig. 23. The extent of the suction zone has been enlarged (from 1.65–1.8 m before the VE to 1.8–2.1 m). There is a large scatter between the predicted suction limit from the different teams (1 m from the tunnel wall up to more than 3 m) although some of the teams succeeded to reproduce the measured profile quite well given the scatter of the data. The partial desaturation limit is not extended, but is strongly intensified. This tendency is well reproduced by the teams.

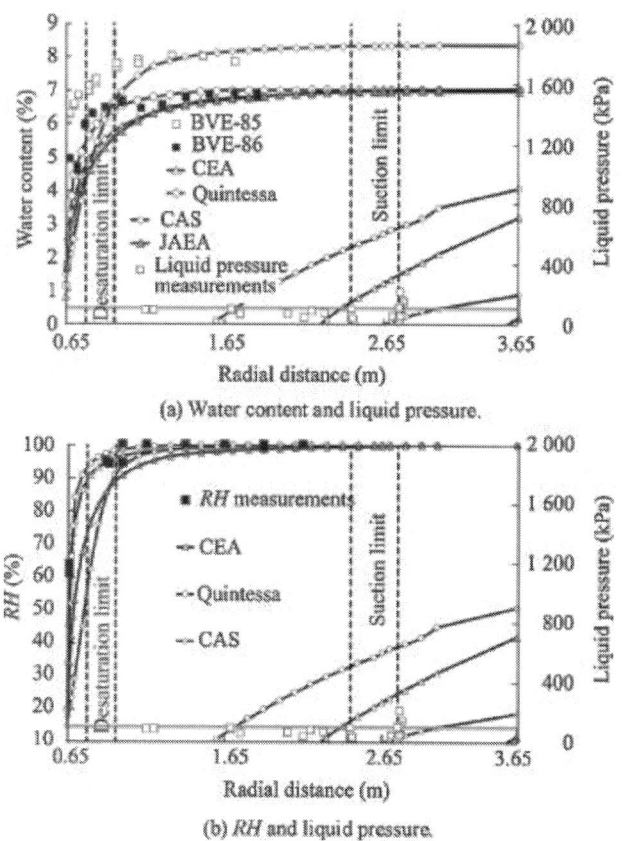

(a) Water content and liquid pressure.

(b) *RH* and liquid pressure.

Figure 23. Measured and simulated profiles of water content, relative humidity and pore water pressure at the end of the first ventilation period (26 January 2004).

Advanced modelling of the VE

After the calibration of the first ventilation phase, minor changes in the parameters set were carried out and new modelling features were introduced. Perhaps, the most significant advance has been the development of models simulating the air flow in the tunnel allowing for prescribing the air flow and RH directly at the in-pipe. These new features were described in details in Bond et al. (2013a). Anisotropic material features were also introduced (Millard et al., 2013). Consideration of water permeability anisotropy explained the differences in water content profiles measured in the bedding plane direction and in perpendicular direction. It also provided a sensible explanation for the scatter observed in the RH and pore water pressure data. The rapid evolution of pore water pressure after ventilation regime changes was qualitatively explained by stiffness anisotropy although the magnitude of the pressure drop could not be reproduced. The largest changes in parameters were introduced by the teams that did not consider vapour diffusion. The influence of vapour diffusion was found less significant than that in the laboratory experiment. The reason is probably that in the in situ test, an inexhaustible water reservoir is available and desaturation is thus less intense than that in laboratory. This caused JAEA and UoE to adopt a permeability value and relative permeability much close to that employed by the teams considering vapour diffusion.

In Phase 1, RH in the rock mass was monitored by hygrometers concentrated between 30 cm up to 2.15 m from the tunnel wall. Before the experiment, the likely position of the saturation limit was estimated to be in that range, but it appeared to be overestimated. Indeed, in that range, all hygrometers provided readings above 95% which is above the confidence limit of these devices. After Phase 1, it was only possible to state that the partially desaturated zone extent was less than 30 cm from the tunnel wall. Four new hygrometers were installed in the first 30 cm before Phase 2. These hygrometers allowed for a better characterization of the RH gradient in Phase 2.

In terms of pore water pressure and determination of the suction limit, the opposite occurred. Most of the piezometers were installed in the suction zone whose extent was estimated to be much close to the desaturation limit before the test. In Phase 2, three new piezometers were installed from gallery 1998 in the farther field of the VE. This allowed a better characterization of the pore water pressure gradient.

The initial conditions before starting the VE are re-examined in Fig. 24. One of the improvements is the better reproduction of the suction limit by all the teams.

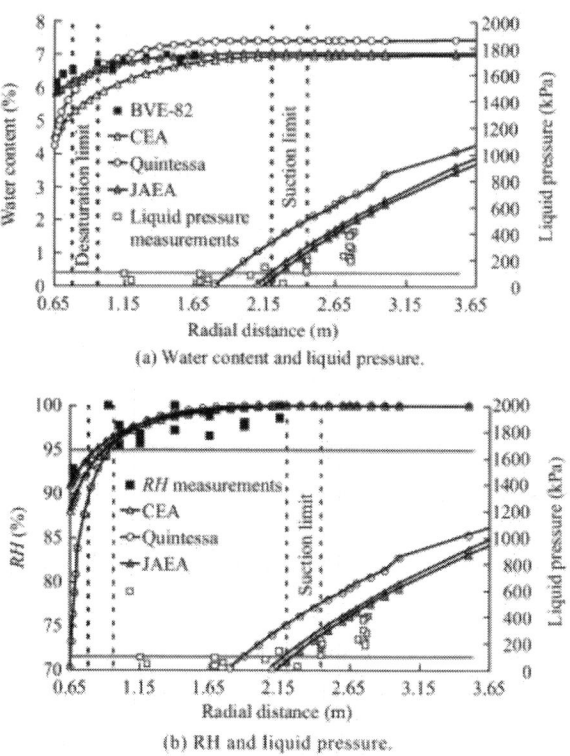

(a) Water content and liquid pressure.

(b) RH and liquid pressure.

Figure 24. Measured and simulated profiles of water content, relative humidity and pore water pressure just before the sealing of the test section (5 July 2002).

The water mass balance is plotted in Fig. 25 over the entire duration of the experiment. It has been reset to 0 at the start of the second phase for comparison purposes. The blind prediction of Phase 2 is almost perfect. JAEA used an improved version of Option 1 for the evaporation boundary condition. They still applied the RH directly, but they used an empirically determined value between the RH measured in the tunnel and that in the first rock centimetres. The calibration of the empirical relationship was done on the basis of Phase 1 and Phase 2 is thus purely predictive. Nevertheless, they still obtained an abrupt response to ventilation regime change. CAS instead changed their boundary condition to Option 2. Quintessa upgraded their model to Option 3 and hence the results are purely predictive. In other words, Quintessa may perform a pure blind prediction of a new VE without any in situ measurement of the new experiment. The RH they computed in the tunnel (Fig. 26) is an output of their modelling approach.

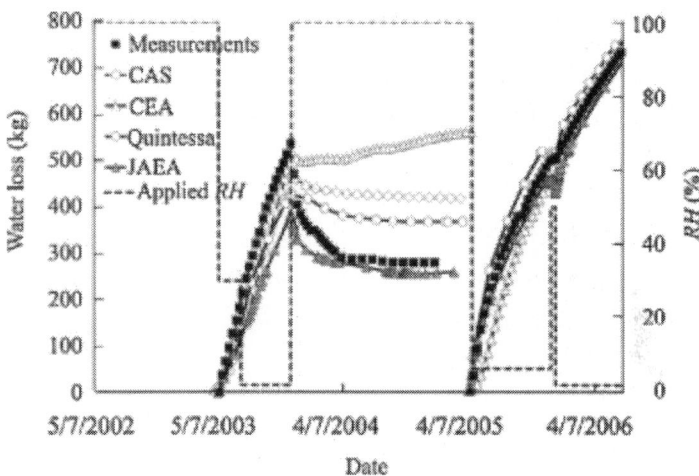

Figure 25. Measured and simulated water mass balance of the test section.

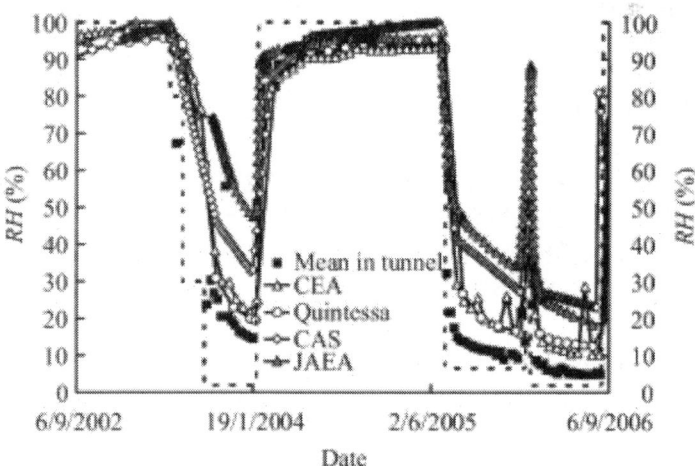

Figure 26. Average relative humidity measured in the test section and calculated or applied relative humidity on the tunnel wall.

The RH evolution in the rock mass is presented in Fig. 27 at different distances from the tunnel wall. The measurements from the newly installed hygrometers in Fig. 27b–d which were not available to the teams during the project provided a good validation test.

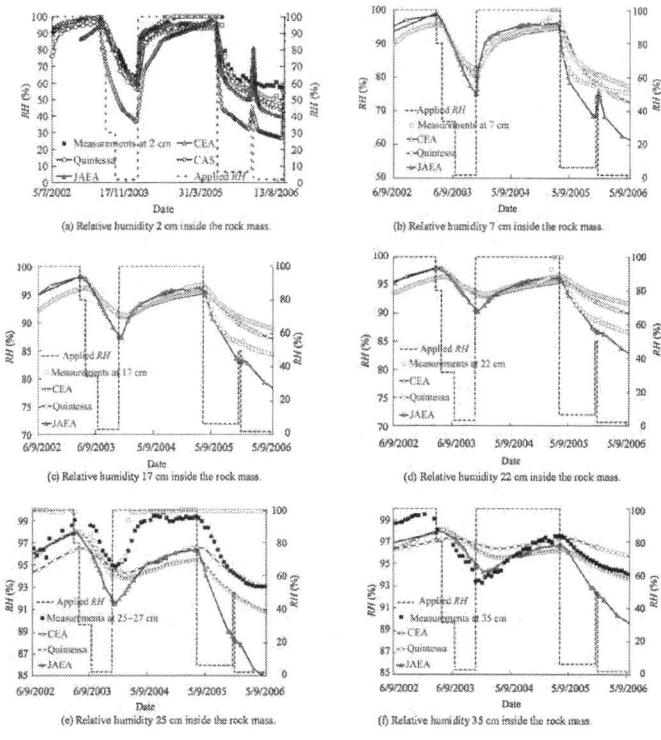

Figure 27. Measured and simulated relative humidity of various points inside the rock mass (advanced modelling).

Reproduction of pore water pressure evolution is only improved by teams that introduced stiffness anisotropy (CEA in Fig. 28). Although the improvement is only qualitative and the magnitude of the pore water pressure drop is not satisfactorily explained, the simulation carried out by CEA allows understanding the process generating the direct pore water pressure response to ventilation (Millard et al., 2013).

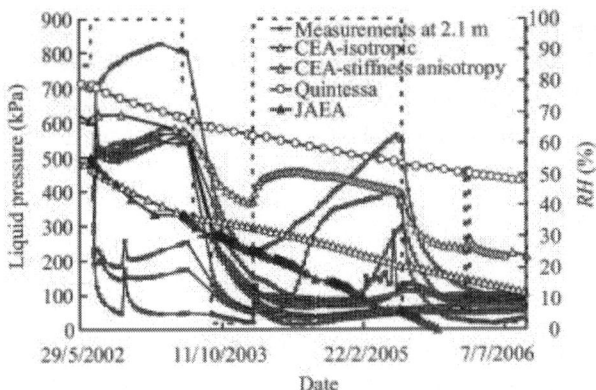

Figure 28. Measured and simulated pore water pressure between 1.8 m and 2.1 m from the microtunnel wall.

The compression–extension behaviour during drying–wetting cycles in the second phase is similar to the one observed in the first phase (Fig. 29). The measurements do not seem to indicate any accumulation of irreversible strains along the cycles although this should be confirmed in another experiment, considering shorter strain measurement intervals (2 m in the VE). Condensation of the ventilation system during the resaturation phase caused flooding of the microtunnel floor. This event seems to have induced an irreversible extension of the rock in the bottom of the microtunnel and even caused failure of one of the extensometers.

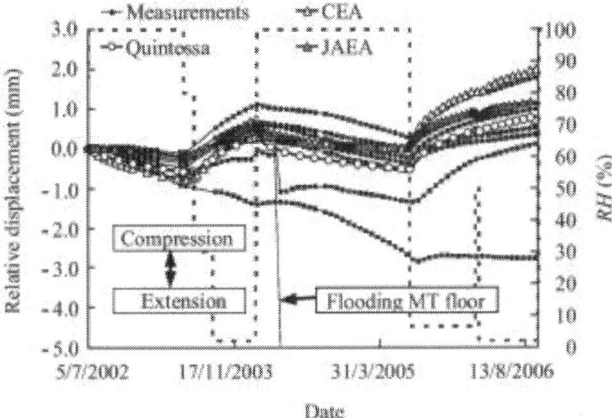

Figure 29. Measured and simulated relative displacement between the microtunnel wall and points 2 m inside the rock mass.

The rock state at the end of the first drying phase is summarized in Fig. 30. According to the measurements carried out in the drilling campaign at the end of the second resaturation phase and the RH measurements, the rock is fully saturated before the start of the second drying phase (Fig. 31). Although some of the farthest piezometers recovered during the resaturation phase, the suction limit did not change significantly. The simulation results reproduced adequately the saturation state before the start of the second drying phase.

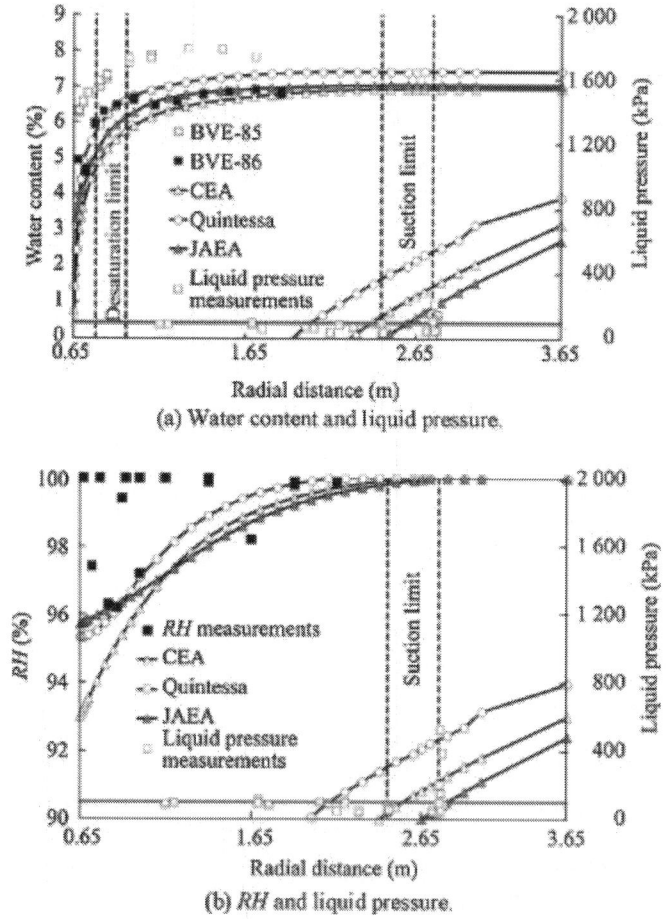

(a) Water content and liquid pressure.

(b) RH and liquid pressure.

Figure 30. Measured and simulated profiles of water content, relative humidity and pore water pressure at the end of the ventilation period (26 January 2004).

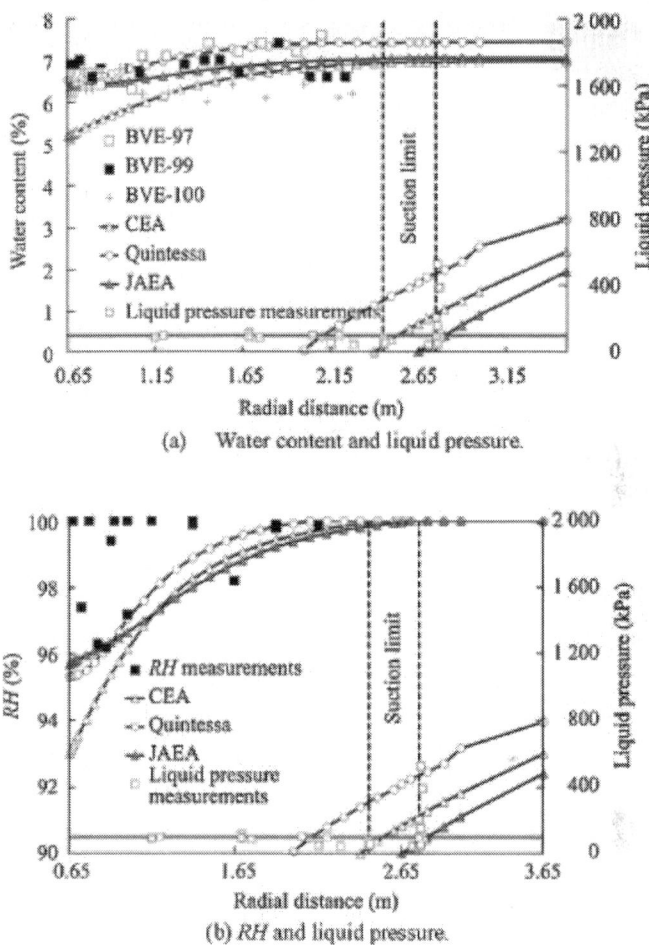

Figure 31. Measured and simulated profiles of water content, relative humidity and pore water pressure at the start of the second resaturation period (1 May 2005).

From the water content and RH measurements, the partially desaturated zone seems to reach 25–30 cm into the tunnel wall at the end of Phase 2 (Fig. 32).

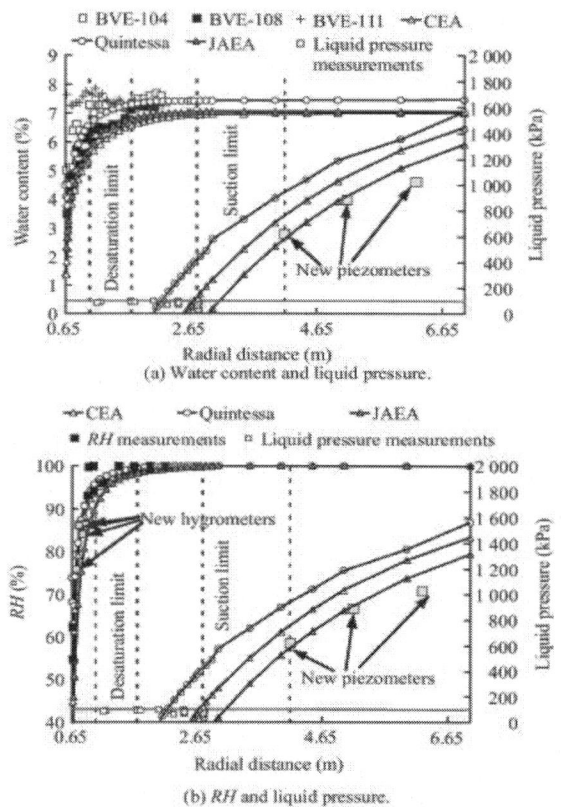

Figure 32. Measured and simulated profiles of water content, relative humidity and pore water pressure at the end of the second ventilation period (4 October 2006).

The measured profiles are quite well reproduced by the different teams. The new hygrometers allowed a more reliable characterization of the RH gradient. Boreholes BVE-104 and BVE-108 were drilled in the test section in the direction perpendicular to the bedding planes and in the bedding plane, respectively. Borehole BVE-111 was drilled between the test section and the tunnel end to evaluate the effect of a year-long closure on the saturation state. All piezometers installed before Phase 1 are in suction, which means that the suction limit is farther than 2.1 m inside the rock mass. The closest new piezometer registered a pore water pressure of about 600 kPa. The suction limit thus certainly does not extent further than 3.5 m, but is likely to be close to 2.1 m, given the high pore water pressure value registered at 3.5 m. The pore water pressure profiles predicted by the teams indicate somewhat higher pressure than the new piezometers. This is probably due to the fact that these piezometers are located in the middle area between the microtunnel and gallery 1998, which was not taken into account in the models.

The long-term evolution of the pore water pressure in the new piezometers is presented in Fig. 33 up to the end of 2010. The measurements exhibit a cyclic behaviour with local maximum each year in February. The cyclic behaviour is related to the temperature variation in gallery 1998 and to the propagation of the heat wave into the rock mass; temperature increases generate pore water pressure increases. Disregarding the thermal cycles, it is possible to state that the long-term prediction provided by the teams is satisfactory given the complexity of the test history and the fact that gallery 1998 has not been taken into account in the modelling.

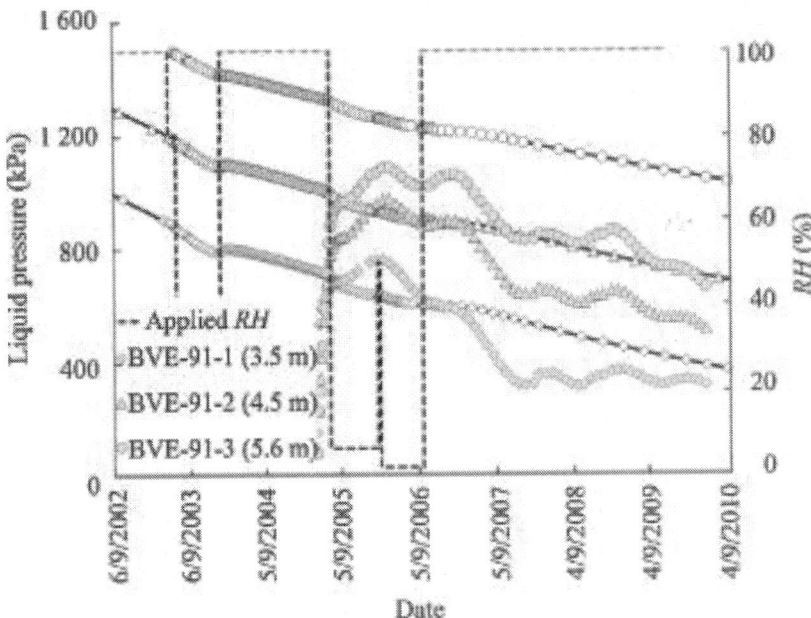

Figure 33. Measured and simulated long-term evolution of pore water pressure between 3.5 and 5.6 m from the microtunnel wall. Simulation results are from CEA.

As a consequence of the marine origin of Opalinus clay, its pore water contains a significant amount of salt. In presence of water evaporation, salt concentration will happen. One of the complementary objectives of the VE was to evaluate the importance of this phenomenon. Comprehensive research has also been carried out on geochemistry processes in the framework of DECOVALEX 2011. Only the main findings are reported here as this is the subject of one of the companion papers (Bond et al., 2013b). First of all, chloride concentration increases were measured near the wall of the tunnel according to the different drying phases although no salt deposition occurred. The chloride gradient could be reproduced by the models using a one-solute transport model as chloride was believed to be a conservative (non-reactive) species. The possible influence of this gradient on the generation of an additional osmotic flow component was assessed and seemed to be totally negligible. Quintessa also run a fully reactive transport model to explain the higher sulphate/chloride concentration ratio in the unsaturated zone than in the saturated zone which indicates the occurrence of chemical reactions.

SUMMARY OF THE IDENTIFIED MECHANISMS/PROCESSES

The analysis of the laboratory drying test and the in situ VE included the identification and discussion of all possible processes that may influence the test results and more particularly the rock behaviour. The modelling work carried out in this project enabled the relevance of a number of desaturation processes to be assessed. A hierarchical classification is proposed:

1) Water transport occurs mainly through water pressure gradients and is thus determined by the permeability of the porous medium.
2) In the unsaturated zone, vapour diffusion has a non-negligible influence. The importance of this process is less in the in situ test than in the laboratory drying test. This is because desaturation is less intense due to an inexhaustible amount of water available in the far field.

3) A correct evaluation of the dependency of the water permeability on the saturation degree was found to be necessary.

4) An accurate representation of the retention curve is necessary, because of its importance in the determination of the permeability of the partially unsaturated zone and of its influence on the boundary between the partial desaturation zone and the suction zone.

5) Anisotropic rock properties (permeability and stiffness) were shown to explain some second order phenomena.

6) Stress redistribution is suspected to have a possible influence on the permeability in the 10–20 cm close to the microtunnel. The good agreement between measurements and simulation results using a permeability value that does not vary with damage indicates that permeability increase in the EDZ may be neglected in experiments involving a similar diameter tunnel.

7) The strains generated by the drying–wetting cycles themselves were very low and did not produce any significant change in porosity, which could have introduced a back influence on the transport problem.

8) Osmotic flow was estimated to have little significance (0.01%– 0.05% of classical Darcy flow).

CONCLUSIONS

The 4-year long controlled ventilation conditions applied to an unlined microtunnel have provided a set of important HM(C) observations. The interpretation of the experiment has been performed with the aid of several numerical tools. Most of the parameters were determined on the basis of a literature review and interpretation of laboratory tests (drying test and retention curve). This approach has allowed the calibration of a number of important rock parameters. The simulations achieved close quantitative agreement with the experiment in many instances and showed their ability to explain qualitatively most observations.

Heavy ventilation conditions applied during 1–2 years were found to produce a partially desaturated zone of quite a limited extent, much smaller than those thought initially. The limit of the partially desaturated zone was estimated to 25–30 cm and the suction limit was about 2 m (or even somewhat more) after a ventilation period of 20 months. The difference between suction and partial desaturation limit has been attributed to the retention behaviour of Opalinus clay.

The VE may be regarded as a huge pump test and may thus be used to determine the hydraulic conductivity of Opalinus clay. Common agreement between the modelling teams seems to indicate an intrinsic permeability value of about 10^{-19} m² for saturated rock. This value lies in the upper range of many measurements carried out at Mont Terri but fits quite well the measurements done in the piezometers of the VE area. The models used led to a very satisfactory reproduction of an important quantity of measurements (water mass balance of the experiment, the relative humidity, pore water pressure, water content and deformation). The consistent results between the different codes and the measurements are an important validation of the numerical tools.

ACKNOWLEDGEMENTS

The work described in this paper was conducted within the context of the international DECOVALEX Project. The authors are grateful to the Funding Organizations who supported the work. The views expressed in the paper are, however, those of the authors and are not necessarily those of the Funding Organizations. The data used in this work were obtained in the framework of the EC project NF-PRO (Contract number FI6W-CT-2003-02389) under the coordination of ENRESA (Empresa Nacional de Residuos Radiactivos).

REFERENCES

1. Bock H. RA experiment rock mechanics analyses and synthesis: data report on rock mechanics. Mont Terri Project: Technical Report 2000-02. Ittigen, Switzerland: Federal Office for Water and Geology (FOWG); 2001.

2. Bond A, Millard A, Nakama S, Zhang C, Garitte B. Approaches for representing hydromechanical coupling between large engineered voids and argillaceous porous media at ventilation experiment, Mont Terri. Journal of Rock Mechanics and Geotechnical Engineering 2013a; 5 (2); in press.

3. Bond A, Benbow S, Wilson J, Millard A, Nakama S, English M, et al. Reactive and nonreactive transport modelling in partially water saturated argillaceous porous media around the ventilation experiment, Mont-Terri. Journal of Rock Mechanics and Geotechnical Engineering 2013b;5(1):44–57.

4. Bossart P, Meier PM, Moeri A, Trick T, Mayor JC. Geological and hydraulic characterisation of the excavation disturbed zone in the Opalinus clay of the Mont Terri rock laboratory. Engineering Geology 2002;66(1/2):19–38.

5. Brooks RH, Corey AT. Properties of porous media affecting fluid flow. Journal of the Irrigation and Drainage Division, ASCE 1966;92(2):61–90.

6. ChijimatsuM, Fujita T, Kobayashi A, NakanoM. Calibration and validation of thermal, hydraulic and mechanical model for buffer material. Tokai, Japan: JNC Technical Report, JNC TW8400 98-017; 1998.

7. Fernández AM, Melón A, Turrero MJ, Villar MV. Geochemical characterisation of the rock samples for the VE-Test before a second cycle of drying. Ventilation test phase II. Madrid: CIEMAT; 2006.

8. Floría E, Sanz FJ, García-Sineriz JL. Drying test: evaporation rate from core samples of ˜ "Opalinus clay" under controlled environmental conditions. Madrid: AITEMIN; 2002.

9. Garitte B, Gens A. The response of an argillaceous rock to ventilation: process identification and analysis of an in situ experiment. In: Qian QH, Zhou YX, editors. Harmonizing rock engineering and the environment, Proceedings of the 12th ISRM international congress on rock mechanics. London: Taylor and Francis Group; 2011. p. 634–5.

10. Garitte B, Gens A, Liu Q, Liu X, Millard A, Bond A, et al. A DECOVALEX-2011 benchmark: laboratory drying test in Opalinus clay. In: EC-TIMODAZ-THERESA THMC Conference. Luxemburg: European Commission; 2009.

11. Garitte B, Gens A, Liu Q, Liu X, Millard A, Bond A, et al. Modelling benchmark of a laboratory drying test in Opalinus Clay. In: Rock Mechanics in Civil and Environmental Engineering. London: Taylor and Francis Group; 2010. p. 767–70.

12. Gens A. HE experiment: complementary rock laboratory tests. Mont Terri Project: Technical Note TN 2000-47; 2000.

13. Gens A, Vaunat J, Garitte B, Wileveau Y. In-situ behaviour of a stiff layered clay subject to thermal loading: observations and interpretation. Geotechnique 2007;57(2):207–28.

14. Gisi M. Evaporation logging FM-D experiment: modification of the equipment. Mont Terri Project: Technical Note 2007-27; 2007.

15. Ippisch O, Vogel HJ, Bastian P. Validity limits for the van Genuchten-Mualem model and implications for parameter estimation and numerical simulation. Advances in Water Resources 2006;29(12):1780–9.

16. Martin CD, Lanyon GW. Measurement of in situ stress in weak rocks at Mont Terri Rock Laboratory, Switzerland. International Journal of Rock Mechanics and Mining Sciences 2003;40(7/8):1077–88.

17. Maul PR. The Quintessa Multiphysics General-Purpose Code QPAC. Warrington, UK: Quintessa Ltd; 2010. Mayor JC, Velasco M. The ventilation experiment phase II (synthesis report). NFPRO Project W.P.4.3; 2008.

18. Meier E. FM-D experiment: evaporation logging in the new gallery. Mont Terri Project: Technical Note 98-51; 1998.

19. Meier E. Evaporation logging (FM-D) experiment, Phase 9: documentation of raw and processed data. Mont Terri Project: Technical Note 2004-50; 2004.

20. Millard A, Bond A, Nakama S, Zhang C, Barnichon JD, Garitte B. Accounting for anisotropic effects in the prediction of the hydro-mechanical response of a ventilated tunnel in an argillaceous rock. Journal of Rock Mechanics and Geotechnical Engineering 2013; 5 (2); in press.

21. Munoz JJ, Lloret A, Alonso E. Characterization of hydraulic properties under satu- ~ rated and non-saturated conditions. NFPRO Project Deliverable 4 EC Contract FIKW-CT2001-00126; 2003.

22. Nussbaum C, Bossart P. Compilation of K-values from packer tests in the Mont Terri rock laboratory. Mont Terri Project: Technical Note 2005-10; 2004.

23. Schaeren G, Norbert J. Tunnel du Mont Terri et du Mont Russelin. La traversée des roches à risques: marnes et marnes à anhydrite, SIA-Dokumentation D037, La traversée du Jura – Les projets des nouveaux tunnels. Zürich: Swiss Society of Engineers and Architects (SIA); 1989.

24. Schuster K. Ventilation test (VE) experiment: final activity report on high resolution seismic investigations within the VE-Experiment. Mont Terri Project: Technical Report 2007-06; 2007.

25. Thury M, Bossart P. The Mont Terri rock laboratory, a new international research project in a Mesozoic shale formation, in Switzerland. Engineering Geology 1999;52(3/4):347–59.

26. Traber D. Geochemical characterisation of samples from drill core BVE82. NFPRO Project Deliverable D5b EC Contract FIKW-CT2001-00126; 2003.

27. Traber D. Geochemical characterisation of samples from drill core BVE85 and BVE86. NFPRO Project Deliverable D5c&d/D22 EC Contract FIKW-CT2001-00126; 2004.

28. Verpeaux P, Millard A, Charras T, Combescure A. A modern approach of large computer codes for structural analysis. In: Hadjian AH, editor. ProcSMiRT. Los Angeles: AASMiRT; 1989. p. 75–85.

29. Villar MV. Retention curves determined on samples taken before the second drying phase. CIEMAT Technical Report M2144/5/07; 2007.

30. Wermeille S, Bossart P. In situ stresses in the Mont Terri Region: data compilation. Mont Terri Project: Technical Report 99-02; 1999.

31. Wileveau Y. THM behaviour of host rock (HE-D) experiment: progress report (Part 1). Mont Terri Project: Technical Report TR 2005-03; 2005.

32. Zhang CL, Rothfuchs T. Report on instrument layout and pre-testing of large lab VE-tests. NFPRO Deliverable 4.3.11; 2005.

CITATION

B. Garitte, A. Bond, A. Millard, C. Zhang, C. Mcdermott, S. Nakama, A. Gens, Analysis of hydro-mechanical processes in a ventilated tunnel in an argillaceous rock on the basis of different modelling approaches, Journal of Rock Mechanics and Geotechnical Engineering, Volume 5, Issue 1, February 2013, Pages 1-17, ISSN 1674-7755, http://dx.doi.org/10.1016/j.jrmge.2012.09.001.

CHAPTER 5

Linked Multicontinuum and Crack Tensor Approach for Modeling of Coupled Geomechanics, Fluid Flow and Transport in Fractured Rock

Jonny Rutqvist[1], Colin Leung[2], Andrew Hoch[3], Yuan Wang[1, 4], Zhen Wang[1, 5]

[1] Lawrence Berkeley National Laboratory (LBNL), Berkeley, CA 94720, USA

[2] Imperial College, London, UK

[3] Serco Energy, Didcot, UK

[4] Hohai University, Nanjing, China

[5] Tongji University, Shanghai, China

ABSTRACT

In this paper, we present a linked multicontinuum and crack tensor approach for modeling of coupled geomechanics, fluid flow, and solute transport in fractured rock. We used the crack tensor approach to calculate effective block-scale properties, including anisotropic permeability and elastic tensors, as well as multicontinuum properties relevant to fracture–matrix interactions and matrix diffusion. In the modeling, we considered stress dependent properties, through stress-induced changes in fracture apertures, to update permeability and elastic tensors. We evaluated the effectiveness and accuracy of our multicontinuum approach by comparing our modeling results with that of three independent discrete fracture network (DFN) models. In two of the three alternative DFN models, solute transport was simulated by particle tracking, an approach very different from the standard solute transport used in our multicontinuum modeling. We compared the results for flow and solute transport through a 20 m × 20 m model domain of fractured rock, including detailed comparison of total flow rate, its distribution, and solute breakthrough curves. In our modeling, we divided the 20 m × 20 m model domain into regular blocks, or continuum elements. We selected a model discretization of 40 × 40 elements (having a side length of 0.5 m) that resulted in a fluid-flow rate equivalent to that of the DFN models. Our simulation results were in reasonably good agreement with the alternative DFN models, for both advective dominated transport (under high hydraulic gradient) and matrix-diffusion retarded transport (under low hydraulic gradient). However, we found pronounced numerical dispersion when using larger grid blocks, a problem that could be remediated by the use of a finer numerical grid resolution, while maintaining a larger grid for evaluation of equivalent properties, i.e. a property grid overlapping the numerical grid. Finally, we encountered some difficulties in using our approach when element sizes were so small that only one or a few fractures intersect an element—this is an area of possible improvement that will be pursued in future research.

INTRODUCTION

In recent years, interest in coupled fluid flow and mechanical processes in fractured rock masses, also known as coupled hydro-mechanical (HM) processes, has increased, along with the ever-increasing demands for energy under sound environmental considerations. The three main fields in which fractured rock coupled HM processes have been most extensively studied are nuclear waste disposal, geothermal energy (hot-dry-rock) extraction, and oil and gas extraction (Rutqvist and Stephansson,

2003). Indeed, coupled HM processes in fractured rock are central to extracting more significant amounts of geothermal energy through enhanced or engineered geothermal systems, as well as to production efforts by the surging North American shale gas industry. Hydraulic stimulation of fractured rock in such systems is used to increase permeability and energy production, but may also induce seismic events that, if felt, can cause concern within the local community. The potential for induced seismicity has also been raised as a concern for deep underground injection and storage of carbon dioxide (Rutqvist, 2012). Under these circumstances, numerical modeling of coupled HM process in fractured rock can be an important tool for finding ways to optimize fluid production or injection, while minimizing the risk of inducing notable seismic events.

Coupled HM numerical models for fractured porous geological media have been available since the early 1980s, including those pioneered by Noorishad et al. (1982), who presented a coupled HM formulation and finite element scheme that later evolved into the computer code ROCMAS. This formulation was based on an extension of Biot's theory of consolidation (Biot, 1941) to include discrete fractures in addition to the porous matrix, and used a fully implicit solution technique. Since then, many computer codes capable of modeling coupled HM processes in fractured porous media have been developed using various numerical methods, including finite element, distinct element and boundary element methods (Jing and Hudson, 2002). At the same time, more realistic constitutive models have been developed to describe coupled HM interaction in rock fractures, with the ones by Barton et al. (1985) and Walsh (1981) most commonly applied (Rutqvist and Stephansson, 2003).

Efforts have also been made to incorporate HM coupling into effective medium theories, such as the crack tensor theory (e.g. Oda, 1986, Stietel et al., 1996, Kobayashi et al., 2001 and Guvanasen and Chan, 2004). The crack tensor is a unique measure combining four significant aspects of hydraulic and mechanical behaviors in fractured rock: volume, fracture size, fracture orientation, and fracture aperture. Oda (1986) applied the crack tensor to HM modeling of rock with geological discontinuities as an anisotropic, elastic porous medium with equivalent elastic compliance and permeability tensors. Stietel et al. (1996) used a similar approach, including two-dimensional (2D) planar analytical treatment of cracks to calculate the equivalent anisotropic HM properties (permeability tensor, stiffness tensor, Biot's coefficient, and Biot's modulus), and compared their results to those from discrete fracture model simulations. Kobayashi et al. (2001)combined crack tensor theory and the Barton–Bandis model for rock joints in analyzing coupled HM processes on shaft sinking in a fractured rock mass. In that study, the elastic stiffness and hydraulic conductivity tensors were updated with resultant anisotropy and heterogeneity during a simulated shaft excavation. Guvanasen and Chan (2004) used what they characterized as a modified crack tensor theory for coupled thermo-hydro-mechanical (THM) processes, which incorporated transformation of a mechanical aperture to a deformed hydraulic aperture for upscaling the THM properties. They applied their model to simulate a large-scale DECOVALEX benchmark test in which the permeability tensor and elastic tensors were upscaled for individual fracture sets, using parameters such as mean length and spacing for each set.

A number of studies have been made in which the effective continuum HM properties have been calculated from numerical simulations using discrete fracture network models (Zhang and Sanderson, 1999, Min et al., 2004 and Blum et al., 2009). For example,Zhang and Sanderson (1999) used such an approach to scale up the 2D conductivity tensor for highly heterogeneous fracture networks. They scaled up local conductivity tensors, based on small sample fracture networks, to estimate the overall nature of conductivity for a relatively large region. Min et al. (2004) used distinct element model to calculate upscaled stress-permeability properties of a fractured rock mass, and then applied it to calculate far-field radionuclide transport under thermally driven stress evolution, also associated with the abovementioned DECOVALEX benchmark test (Min et al., 2005). Blum et al. (2009) also applied their modeling approach to the same benchmark test, but used a stochastic analysis for investigating the impact of various basic HM properties on the transport simulation results. The spread of travel times in the simulation results demonstrated the significance of both HM rock properties and their spatial distribution.

For a fractured porous medium, it might be useful to represent the fractures and intervening rock matrix by overlapping fracture and matrix continua, referred to as dual-continuum or dual-porosity models. This is a common approach in reservoir engineering where the storage of fluids might take place primarily in the matrix, with permeability controlled by fluid flow in fractures (e.g. Barenblatt et al., 1960 and Warren and Root, 1963). A generalization of this concept is realized by the multiple interacting continua or MINC model, a concept that has been extensively applied in multiphase flow and transport modeling (Pruess and Narasimhan, 1985). Studies incorporating both Biot's poroelastic theory and dual-porosity concept include work on single-phase fluid flow byBerryman and Wang (1995), Chen and Teufel (1997), Bai (1999), and Berryman (2002), as well as recent developments associated with multiphase fluid flow by Liu and Rutqvist (2012) and Kim et al. (2012). A version of the dual-continuum concept, the dual-permeability concept, was applied for modeling coupled THM processes associated with the Yucca Mountain Project, Nevada (Rutqvist and Tsang, 2003). This included modeling of the Yucca Mountain drift scale test (Rutqvist et al., 2005 and Rutqvist et al., 2008) and also part of the previous DECOVALEX studies (Rutqvist et al., 2009a). In these modeling efforts, the emphasis was on high-temperature (above boiling for water) thermally driven coupled processes in the near-field of heated drifts, including heat advection in the fracture system and dryout of matrix rock.

In this paper, we present a linked multicontinuum and crack tensor approach and apply them to a simulation case involving coupled geomechanics, fluid flow, and solute transport in fractured rock. We apply the crack tensor theory to calculate effective block scale properties, including anisotropic permeability and elastic tensors, as well as multicontinuum properties relevant to matrix diffusion. Stress-dependent properties are considered through stress-induced changes in fracture apertures, which in turn are used to update permeability and elastic tensors as a function of mechanical loading. We test the approach in a simulation study of water flow and solute transport through a 20 m × 20 m model domain of fractured rock subjected to changing stress conditions (Fig. 1 and Fig. 2). This simulation case was a part of the international DECOVALEX-2011 project, where it was denoted Task C (Zhao et al., 2013). The rock mass considered in this example is heavily fractured; the model domain contains 7797 individual fractures, with an average fracture spacing of about 0.13 m. The geometric fracture data originate from the Sellafield area, West Cumbria, UK, and have been used as input for several previous DECOVALEX studies (Kobayashi et al., 2001, Blum et al., 2009, Min et al., 2004 and Öhman and Niemi, 2003). Four fracture sets are identified having fracture length following a power law with fitted fractal dimension equal to 2.2 (Min et al., 2004). The fracture network was generated for minimum and maximum cutoffs of 0.5 m and 250 m trace length, respectively. Despite a minimum cutoff trace length of 0.5 m, the mean trace length is still just 0.92 m, and more than 95% of the fractures are less than 2 m (Min et al., 2004). Baghbanan and Jing (2007) used the same statistical fracture parameters, but generated a fracture network that included a correlation between fracture aperture and length, so that larger fractures would have larger apertures and higher permeability. The aperture–length correlation used, shown in Fig. 3, has a profound effect on the fluid flows distribution, with more fluid going through larger fractures, leading to a more heterogeneous flow field (Baghbanan and Jing, 2007).

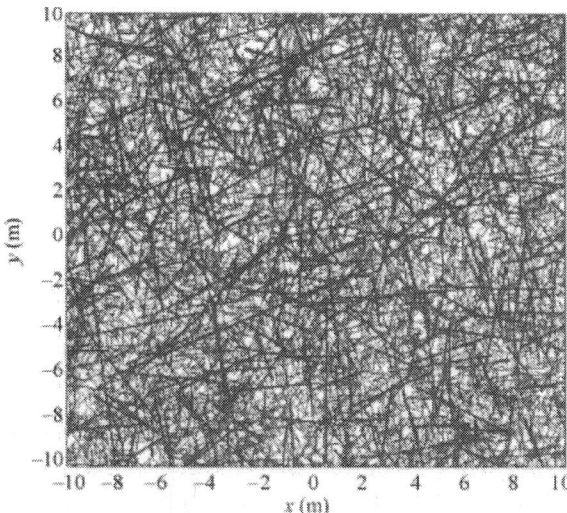

Figure 1. A 20 m × 20 m model domain and the 2D fracture network with 7797 fractures.

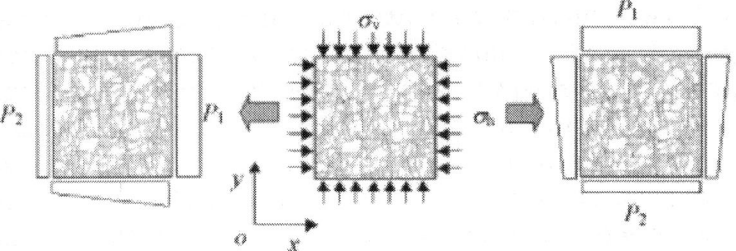

Figure 2. Mechanical and hydraulic boundary conditions for investigating flow and transport under mechanical load. P_1 and P_2 are the upstream and downstream boundary fluid pressures, respectively, and σ_v and σ_h are the vertical and horizontal boundary stresses, respectively.

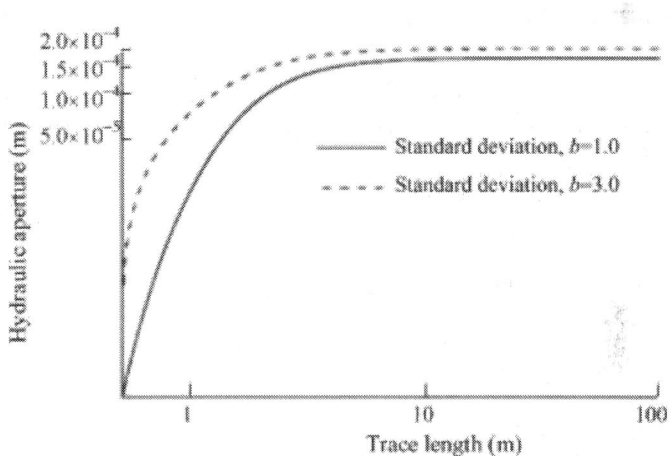

Figure 3. Hydraulic aperture correlation with fracture trace length according to Baghbanan and Jing (2007). In this case, the solid line correlation function for $b = 1.0$ was used.

In this study, we evaluate the effectiveness and accuracy of our linked multicontinuum and crack tensor approach, by comparing our modeling results with that of three independent discrete fracture network (DFN) models applied to the same benchmark problem within the DECOVALEX-2011 Project (Zhao et al., 2013). In two of the three alternative DFN models, solute transport was simulated by particle tracking — very different from our standard solute transport model. We compared the results for fluid flow and solute transport through the entire 20 m × 20 m model domain, with detailed comparison of total flow rate, its distribution, and breakthrough curves for flow and transport under both high and low pressure gradients, respectively, with low and high impacts on fracture-to-matrix diffusion. It is shown that the multicontinuum approach is indeed a viable option to DFN models for modeling of coupled fluid flow, solute transport, and geomechanics in fractured rock masses, although we have identified some needed improvements that will be the subject of future research.

LINKED MULTICONTINUUM AND CRACK TENSOR APPROACH

In this section, we describe the approach, implementation, and calculation procedure for the linked multicontinuum and crack tensor approach. We implemented it in TOUGH-FLAC (Rutqvist et al., 2002 and Rutqvist, 2011), a simulator based on coupling the TOUGH2 multiphase flow simulator (Pruess et al., 1999) with the FLAC3Dgeomechanical simulator (Itasca, 2009). We also employed the TOUGH2 multiple interaction continuum (MINC) approach (Fig. 4), which provided us with an effective way of including important fracture–matrix interactions, and diffusion.

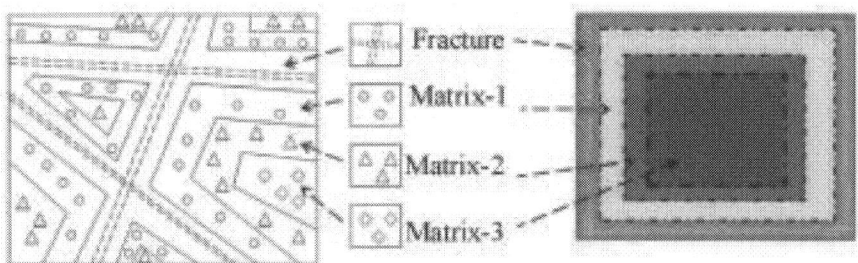

Figure 4. Schematic diagrams for the multiple porosity model and MINC concept according to Pruess and Narasimhan (1985). Left: a composite porous medium which consists of several distinct types of porous materials. Right: a conceptual diagram for a MINC model (Kim et al., 2012).

General Approach and Domain Discretization

In the linked multicontinuum and crack tensor approach, we divide the model domain into overlapping fracture and matrix continuum elements (Fig. 5). For each element, we calculate upscaled (effective) properties, such as permeability and elastic tensors. We also calculate multicontinuum parameters, such as average fracture spacing and fracture volume fraction that are used in the TOUGH2-MINC model. For example, in the DECOVALEX-2011 benchmark test, we divided the 20 m × 20 m fractured rock domain into uniform cubic elements. We investigated the use of different element sizes, from a very fine mesh of 400 × 400 elements (i.e. elements with side length of 0.05 m) to a very coarse mesh of 4 m × 4 m (i.e. side length of 5 m). In the case of a very fine mesh (elements with side length of 0.05 m), only 1 or 2 fractures, or indeed none may intersect each element. The fine-mesh approach could be characterized as a discrete continuum approach, because each fracture trace is actually discretely represented through distinctly different properties of grid elements along a fracture trace. Such an approach was previously introduced and applied using TOUGH-FLAC by Rutqvist et al. (2009b) for studies related to the excavation disturbed zone (EDZ) around tunnels in fractured rock. However, herein we use a more systematic approach, in which equivalent properties (such as permeability and elastic tensor) for each fracture continuum element (Fig. 5) are calculated analytically, using crack tensor theory.

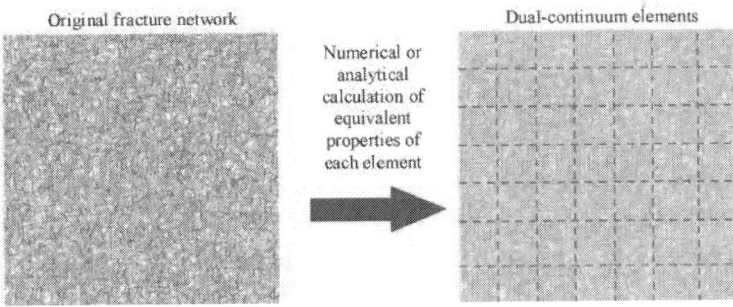

Original fracture network

Numerical or analytical calculation of equivalent properties of each element

Dual-continuum elements

Figure 5. Scheme of multicontinuum approach used in this study. The original fracture network was simulated using overlapping fracture and matrix continua with equivalent properties of the fractured continuum elements calculated using Oda's crack tensor theory, for comparison purpose the equivalent permeability was also calculated numerically using DFN modeling with NAPSAC.

Effective multicontinuum properties using crack tensor theory

The effective multicontinuum properties were derived using Oda's crack tensor theory (Oda, 1986), but in this case formulated as a discrete summation of contributions from each fracture that intersects an element volume. We can use discrete summation for this case, because each fracture is known by its position and its geometric properties — length, orientation, and aperture. We use Oda's crack tensor theory, in which fractures are considered as disc-shaped in a 3D system. The basic quantities of the crack tensor for each crack intersecting an element can be stated as follows:

$$F_{ij} = \frac{1}{V_e} \frac{\pi}{4} D^3 n_i n_j \tag{1}$$

$$F_{ijkl} = \frac{1}{V_e} \frac{\pi}{4} D^3 n_i n_j n_k n_l \tag{2}$$

$$P_{ij} = \frac{1}{V_e} \frac{\pi}{4} D^2 b^3 n_i n_j \tag{3}$$

where F_{ij}, F_{ijkl}, P_{ij} are the basic crack tensors; V_e is the element volume; D is the diameter of the crack; b is the aperture of the crack; and n is the unit vector of normal orientation for each crack with the component n_i $(i = 1, 2, 3)$.

By using the quantities and the mechanical properties for each fracture, we calculate the anisotropic compliance tensor C_{ijkl} and permeability tensor k_{ij}, using

$$C_{ijkl} = \sum^{NCR} \left[\left(\frac{1}{K_n D} - \frac{1}{K_s D} \right) F_{ijkl} \right.$$
$$\left. + \frac{1}{4 K_s D} (\delta_{ik} F_{jl} + \delta_{jk} F_{il} + \delta_{il} F_{jk} + \delta_{jl} F_{ik}) \right] \tag{4}$$

$$k_{ij} = \sum^{NCR} \frac{1}{12}(P_{kk}\delta_{ij} - P_{ij}) \tag{5}$$

where NCR is the number of fractures (or cracks) intersecting an element, K_n is the fracture normal stiffness, K_s is the fracture shear stiffness, and δjl is the Kronecker's delta.

The total elastic compliance tensor can be then be formulated as

$$T_{ijkl} = C_{ijkl} + M_{ijkl} \tag{6}$$

$$M_{ijkl} = \left(\frac{1}{E}\right)[(1+v)\delta_{ik}\delta_{jl} - v\delta_{ij}\delta_{kl}] \tag{7}$$

where $Mijkl$ is the elastic compliance tensor of the intact rock that depends on the Young's modulus, E, and the Poisson's ratio, v.

For a multicontinuum model, we need to calculate the fracture volume fraction in each element. However, in the current model simulation, and in the MINC capability of TOUGH2, the fracture volume fraction, f_{vf}, is derived for the entire domain, as follows:

$$f_{vf} = \frac{\sum^{NE}\sum^{NCR}(1/4)\pi D^2 b}{\sum^{NE} V_e} \tag{8}$$

where NE is the number of elements. In the current application for the DECOVALEX-2011 benchmark test, we must consider that the problem was defined as a 2D problem, with fractures defined by their end points and aperture. When using Oda's crack tensor for this 2D system, we divide the domain into cubic blocks and investigate the geometry of each fracture intersecting such a block. The intersection of a fracture plane with an element volume is defined as a polygon with segments. We then calculate an equivalent disc-shaped fracture diameter, preserving a consistent fracture area (and fracture volume). Having the equivalent diameter of the disc-shaped fracture, we can readily apply Oda's theory.

Stress effects on permeability and compliance tensors
The effects of stress on permeability and compliance are considered through constitutive models for hydraulic and mechanical behaviors of single fractures under stress. In this case, we use a fracture HM constitutive model defined in the description of the benchmark test. It was derived as a simplified version of the Bandis hyperbolic model (Bandis et al., 1983), constrained by some basic relationships between initial aperture, maximum normal closure, and stiffness that could represent the observed sample size dependency of these parameters. For example, Baghbanan and Jing (2008) defined a normalized critical normal stress, as the fracture is compressed and approaches the maximum closure where normal compliance decreases significantly. For their particular study, which is also used in this study, the normalized critical stress was set to 10. Moreover, they assumed that maximum normal closure δ_m over initial (physical) aperture h_i is constant and equal to 0.9, i.e. $\delta_m/h_i = 0.9$. Using these assumptions, Baghbanan and Jing (2008) simplified the Bandis hyperbolic normal closure equation as follows:

$$\sigma_n = \frac{\sigma_{nc}\delta}{10(0.9h_i - \delta)} \tag{9}$$

and then normal stiffness can be linked to the normal stress as given by

$$K_n = \frac{(10\sigma_n + \sigma_{nc})^2}{9\sigma_{nc}h_i} \tag{10}$$

where σ_{nc} is the critical normal stress (MPa), defined by Baghbanan and Jing (2008) as $\sigma_{nc} = 0.487h_i + 2.51$. This relation for critical normal stress was chosen such that critical normal stress would vary linearly from 3 to 100 MPa for the defined range of initial aperture (from 1 μm to 200 μm).

In the benchmark test description, shear slip was to be modeled using an elasto-perfectly plastic model with a Coulomb failure criterion, including a constant shear stiffness $K_s = 434$ GPa/m (Zhao et al., 2013). As soon as shear failure is reached, the continued shear displacement, u_s, would induce shear dilation, u_{dil}, according to

$$u_{dil} = u_s \tan \varphi_d \qquad (11)$$

where φ_d is the dilation angle, which in this study was set to 5°. In the Coulomb failure criterion, the cohesion $c = 0$, and the internal friction angle is set to $\varphi = 24.9°$, a value once determined from laboratory tests on core samples related to fractures from one of the fracture sets.

In our model simulations, we then consider stress-induced aperture change as a result of both current normal stress and shear dilation as follows:

$$b = b_0 - \delta + \Delta b_{dil} \qquad (12)$$

$$\delta = \frac{9\sigma_n b_0}{\sigma_{nc} + 10\sigma_n} \qquad (13)$$

where δ is the normal closure caused by an increase in normal stress, and Δb_{dil} is the dilatational change in aperture taken as equal to the shear dilation, i.e. $\Delta b_{dil} = u_{dil}$. Eq.(13) was derived from Eq. (8), denoting the initial aperture as b_0, i.e. $b_0 = h_i$. Note that $b_0 = h_i$ is the aperture under stress-free conditions that might be measured on core samples, but in this study, $b_0 = h_i$ was assigned for each fracture based on the aperture–length correlation shown in Fig. 2.

In this study, we propose a pragmatic approach for modeling of fracture shear slip and dilation, by considering the shear stiffness of the rock surrounding the fracture being sheared. We consider the effect of shear failure through reduction in the shear stiffness, and that shear dilation starts at the critical shear stress τ_{sc} dictated by the Coulomb criterion. A useful feature in this approach is that the fracture shear displacement remains a unique function of the current shear stress even after failure. Thus, shear dilation can be directly related to the current shear stress across the fracture.

Fig. 6 illustrates how the shear stiffness of the system — defined as $K_s = K_s^{fracture} + K_s^{rock}$ — changes at the onset of shear failure. $K_s^{fracture}$ is the shear stiffness related to interlocking of the two opposite rough-fracture surfaces, and it is given in the benchmark test description as 434 GPa. K_s^{rock} is the shear stiffness of the surrounding rock mass felt by the fracture, estimated from an expression related to shear stress drop, $\Delta\tau$, and shear displacement, u_s, along a fracture embedded in a linear-elastic medium as (Dieterich, 1992):

$$K_s^{\text{rock}} = \frac{\Delta\tau}{u_s} = \eta\frac{G}{l}$$

(14)

where G is the shear modulus of the surrounding rock mass, l is the half length (or radius r) of the fracture, and η is a factor with a value that depends upon the geometry of the slip patch and assumptions related to slip or stress conditions on the patch (Dieterich, 1992). In our case, we may consider a circular crack in which $\eta = 7\pi/24 \approx 0.92$ and l is the radius r. For example, in this benchmark test study, the elastic matrix properties given in Table 1 yield a shear modulus, $G = 34.1$ GPa. Shear stiffness of the rock would then be

$$K_s^{\text{rock}} = \eta\frac{G}{l} = \frac{7\pi}{24}\frac{G}{r} \approx \frac{31.2 \times 10^9}{r}$$

(15)

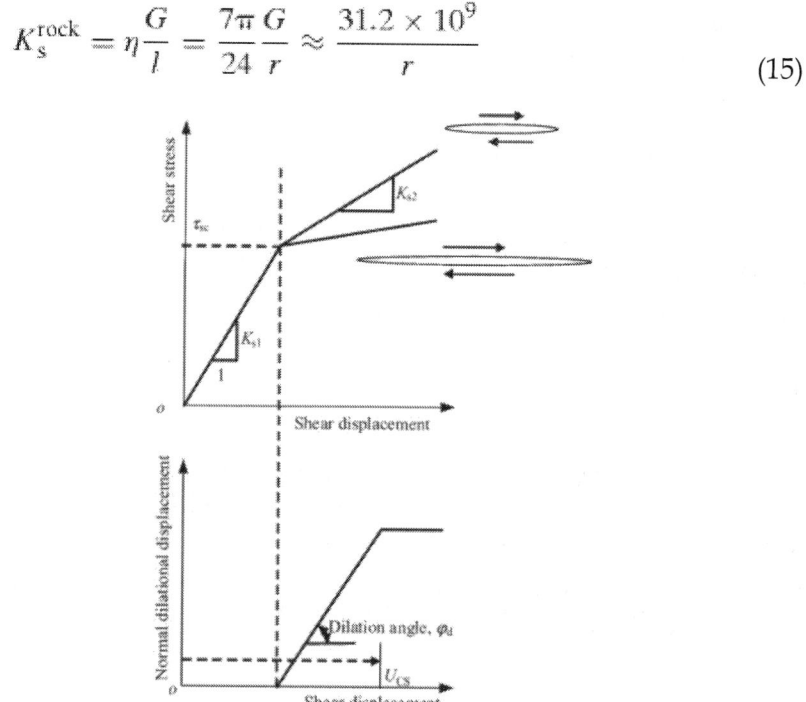

Figure 6. Approach for modeling of dilatational normal displacement as a result of shear slip in fractures embedded in a rock mass.

Table 1. Model properties of intact rock and fractures.

Intact rock		Fractures						
Elastic modulus, E (GPa)	Poisson's ratio, ν	Shear stiffness, K_s(GPa/m)	Friction angle,φ (°)	Dilation angle,φ_d (°)	Cohesion, c (MPa)	Critical shear displacement for dilation, u_{cs} (mm)	Minimum aperture value, a_{res} (μm)	Maximum aperture value, a_{max} (μm)
84.6	0.24	434	24.9	5	0	3	1	200

For a fracture length of 0.92 m (the mean fracture trace length in this study), $K_s^{rock} = 67.8\,\text{GPa/m}$, i.e. much less than $K_s^{fracture} = 434\,\text{GPa/m}$. Before shear failure, the system stiffness will then be $K_s = K_s^{fracture} + K_s^{rock} = 434 + 67.8 = 501.8\,\text{GPa/m}$. After shear failure, the elasto-perfectly plastic assumption corresponds to $K_s^{fracture} = 0$, and hence $K_s = K_s^{rock} = 67.8\,\text{GPa/m}$, given by Eq. (15). Thus, in this case, shear stiffness drops significantly upon shear failure. Fig. 6 also illustrates that longer fractures result in lower K_s^{rock}, which is also obvious from Eq. (14).

For implementation into the TOUGH-FLAC simulation, we take advantage of the fact that shear dilation occurs for shear stress changes happening after the onset of shear slip, and therefore the dilatational normal displacement can be calculated using

$$\Delta b_{dil} = \frac{\tau - \tau_{sc}}{K_s^{rock}} \tan \varphi_d \qquad (16)$$

In our TOUGH-FLAC model simulation, we calculate a stress-induced (including normal and shear stresses) aperture update to a new aperture and fracture normal stiffness, and use these new values in the calculation of updated permeability and elastic tensors. The workflow for doing this is described in the next section.

Implementation and work flow

Fig. 7 presents the calculation procedure for the linked multicontinuum and crack tensor approach. We first calculate the initial (stress free) elastic compliance tensor and permeability tensor, considering fracture geometry and mechanical properties. For a stress ratio $K = 1$, we first assign initial total elastic compliance tensor fields to each element in the FLAC3D model (employing a user-defined elastic anisotropic constitutive model) and calculate the new stress state. Note that, because of the heterogeneous compliance within the model, the new stress state will be heterogeneous, somewhat different from the stress applied at boundaries. Then, we use the new stress fields to calculate the stress induced fracture change and update the permeability and compliance tensor fields. With the permeability tensor defined in each element, we use TOUGH2 to simulate the flow and transport through the model. The same procedures are used for stress ratios $K = 2, 3$, and 5.

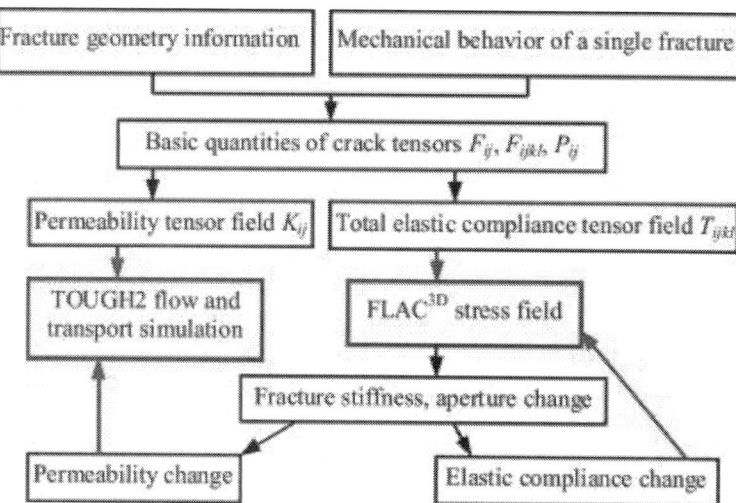

Figure 7. Workflow of effective property calculation using crack tensor.

MODEL SETUP

We construct the model for testing and comparison to alternative DFN models for fluid flow and transport in fractured rock under mechanical loading. As mentioned in our approach, we use a standard solute transport calculation rather than particle tracking, and we also use multiple continua, including fracture and matrix continua. In this section, we describe how data and tasks defined in the benchmark description were adapted for the current multicontinuum approach.

Multicontinuum material properties

The basic parameters for matrix and fractures are given in Table 1, whereas parameters related to fracture–matrix interaction and diffusion are given in Table 2. Notably, only elastic parameters with no permeability were defined for the matrix, which in such a case would be linearly elastic and impermeable. These are the material parameters used in the UDEC simulations of this fracture network in Baghbanan and Jing (2008). For our multicontinuum model, we assign a matrix permeability of 1×10^{-20} m², which is so small that matrix flow is insignificant, compared to the total rock mass flow. Overall, the permeability of the fractured rock mass in this case is approximately 1×10^{-13} m². Such a high permeability would not be realistic for sites in competent granitic rocks, but was encountered in the volcanic tuff of the Yucca Mountain site in Nevada, which is also intensively fractured (Rutqvist et al., 2008 and Rutqvist and Tsang, 2003).

Table 2. Parameters for multicontinuum model.

K	Fracture			Matrix			Molecular diffusion coefficient (m²/s)
	Porosity	Tortuosity	Fracture spacing (m⁻¹)	Fracture volume fraction	Effective transport porosity, n_e	Tortuosity, τ_0	
0	1	1.05×10^{-3}	0.2	1.05×10^{-3}	3.16×10^{-3}	1×10^{-2}	1×10^{-9}
1	1	0.61×10^{-3}	0.2	0.61×10^{-3}	3.16×10^{-3}	1×10^{-2}	1×10^{-9}
3	1	0.46×10^{-3}	0.2	0.46×10^{-3}	3.16×10^{-3}	1×10^{-2}	1×10^{-9}
5	1	0.41×10^{-3}	0.2	0.41×10^{-3}	3.16×10^{-3}	1×10^{-2}	1×10^{-9}

In the TOUGH2-MINC model, we applied both dual-continuum (one fracture and one matrix continuum) and multicontinuum (one fracture and three matrix continua) approaches, with the multicontinuum approach achieving better results, because steep gradients between fracture and matrix could be better resolved. The input to TOUGH2-MINC includes fracture volume fraction and average fracture spacing, used for internal mesh properties (such as element volumes, connection distances, and interface areas). The fracture volume fraction was calculated using Eq. (8) for each loading case, resulting in a decreasing fracture volume fraction with increasing load. The average fracture spacing can be calculated from the inverse of $P32$, which is defined as the fracture surface area per unit volume, leading to an average fracture spacing of $1/P32 \approx 0.13$ m. However, in the simulations, as listed in Table 2, an average fracture spacing of 0.2 m was assumed to be consistent with some of the other teams who assumed a matrix diffusion distance of 0.1 m.

Applied mechanical loading and pressure gradients
According to the benchmark test description, various boundary stresses were applied on the model domain to generate deformed states for the fluid flow and transport analyses (Fig. 1 and Fig. 2). A constant vertical normal compressive stress of 5 MPa was specified at the top and bottom boundaries, and a horizontal normal compressive stress, varying from 5, 10, 15 to 25 MPa, was applied at the left and right boundaries. In this way, the stress ratio, defined as horizontal/vertical stresses, increased from $K = 0$, 1, 2, 3 to 5 stepwise ($K = 0$ represents the free state when both horizontal and vertical stresses are zero). Thus, the shear stress and therefore the potential for inducing shear slip along fractures will increase with each loading step.

Fluid flow through the model under various stress ratios was simulated under the specified hydraulic pressure gradients illustrated in Fig. 2b and c. The hydraulic boundary conditions allowed the fluid and solutes to exit from three outlet boundaries, except in some cases where closed (impermeable) lateral boundaries were assigned and the solute could exit only through the downstream boundary. Two sets of hydraulic conditions were defined to obtain the macroscopic flow in vertical and horizontal directions, (Fig. 2b and c), respectively. Moreover, two different values of hydraulic gradient were applied. Initially, a gradient of 10 kPa/m was defined, leading to fast fracture flow, in which the solute transport and breakthrough curves basically depended on the distribution of advective flow within the fracture system, without any significant effect on matrix diffusion. Actually, a gradient of 10 kPa/m is much higher than the conditions expected at a site for an underground radioactive waste repository. Therefore, an additional case of applying a more realistic gradient of 10 Pa/m was defined with matrix diffusion significantly affecting the transport.

Solute injection and monitoring

In our model simulations, using the linked multicontinuum and crack tensor approach, we used a standard solute transport model which is part of the TOUGH2 code and applied the TOUGH2/EOS1 equation of state module (Pruess et al., 1999). In the benchmark for the DECOVALEX-2011 Project, the solute transport was described in terms of particle tracking (Zhao et al., 2013). Solute particles were to be introduced at an amount proportion to the flow rate at each inlet location. Particles collected at each outlet boundary as a function of time were presented in the form of breakthrough curves (Zhao et al., 2013). In our standard solute-transport approach, we used TOUGH2/EOS1 and injected a second water component (water component 2) at the inlet boundary in a pulse over a short time period (e.g. 10 s). We then monitored the mass flow of water component 2 at each outlet boundary to calculate breakthrough curves.

Matrix diffusion

In the benchmark test description, a matrix pore diffusion coefficient, $D_p = 10^{-11}$ m^2/s, was given as well as a matrix porosity of 0.136%, whereas effective diffusion used in the TOUGH2 continuum model was defined as $D_e = D_p n_e$, where n_e is the effective transport porosity (Neretnieks, 1980). Specifically, in TOUGH2, the diffusive mass flux componentκ in phase β is calculated as

$$ f_\beta^\kappa = -n_e \tau_0 \tau_\beta \rho_\beta d_\beta^\kappa \nabla X_\beta^\kappa \qquad (17) $$

where $\tau_0 \tau_\beta$ is the tortuosity, which includes a porous medium dependent factor τ_0 and a coefficient, τ_β, that depends on the phase saturation; ρ_β is the density; d_β^κ is the diffusion coefficient of component κ in phase β ; and ∇X_β^κ is the gradient of the mass fraction of component κ in phase β. In this case, we have a single phase (liquid), with two water components (water 1 and water 2), in which water 2 corresponds to the injected solute. Therefore, $\tau\beta = 1$, and the effective diffusion and pore diffusion coefficient would correspond to $D_e = n_e \tau_0 d_\beta^\kappa$ and $D_p = \tau_0 d_\beta^\kappa$, respectively. We interpret this such that within the matrix continuum, the molecular diffusion coefficient for the tracer (or solute) in water would be 10^{-9} m²/s, with a tortuosity factor $\tau_0 = 0.01$, whereas in the fracture continuum, the fracture porosity is set to 1, and the tortuosity factor is equal to the fracture volume fraction calculated from Eq. (8). As mentioned in our current model, we assign a homogenous fracture volume fraction in the entire model domain.

RESULTS AND COMPARISON

Here we present the modeling results for flow and transport through the 20 m × 20 m model domain under increasing boundary stresses. We focus on the linked multicontinuum and crack tensor results denoted by LBNL, but also compare with results from the alternative DFN model simulations. The DFN model simulations were conducted by research teams from Imperial College (IC), UK, using the NAPSAC DFN model; the Royal Institute of Technology (KTH), Sweden, using distinct element method simulations with UDEC; and the Technical University of Liberec (TUL), using their finite element method (FEM) code Flow123 (Zhao et al., 2013). The IC and KTH teams used particle tracking, whereas the TUL team used a standard solute-transport model, equivalent to the one used in our model. Further details about the DFN models and results can be found in Zhao et al. (2013).

Fluid flow without mechanical loading

A great deal of effort was spent on development and testing of the algorithms for calculating the permeability tensor using Oda's theory. The permeability field developed by the Oda's crack tensor was compared to that calculated by an alternative numerical DFN flow calculation according to Jackson et al. (2000), using the NAPSAC DFN code. Moreover, we also compared our TOUGH2 finite volume flow simulations with that of an alternative FEM simulation, using the NAMMU code (Jackson et al., 2000). We investigated the effect of dividing the 20 m × 20 m model into 400 × 400, 100 × 100, 40 × 40, 10 × 10, and 4 × 4 elements. Thus, the size or side length of these elements is 0.05 m, 0.2 m, 0.5 m, 2 m, and 5 m, respectively. The results of this comparison indicated some problems when using Oda's crack tensor theory and TOUGH2 finite volume scheme as is in particular when element sizes became small compared to the density of the fracture network.

Fig. 8 presents an example of vertical flow distributions using a TOUGH2 simulation and a 400 × 400 mesh for the numerically evaluated (using NAPSAC) and analytically evaluated (using Oda's theory) permeability tensor. The black lines inside the model have a width proportional to the flow rate; outflow distribution is also shown. The TOUGH2 model simulation provides a flow distribution similar to the DFN modeling, with flow dominant along larger fractures, but with the total outflow only amounting to about 70% of the DFN model results. The problem was identified to be associated with (1) element size smaller than a representative elementary volume, and/or (2) finite difference method (FDM) space discretization approximations in TOUGH2. The second item becomes more apparent when imagining the case of only one fracture intersecting an element. If it is a diagonal fracture, and if TOUGH2 uses a standard 5-point finite difference approximation, there are only vertical and horizontal connections between numerical elements, and the permeability along each such connection is projected from the permeability tensor at each element. In the case of a diagonal fracture, the fluid must flow a longer path in the 5-point differencing (Fig. 9a). We tried to remedy this problem by using 9-point differencing (Pruess, 1991) that allows flow between diagonal-element connections (Fig. 9b). However, for a very fine mesh, the flow rate is still low compared to the DFN model results.

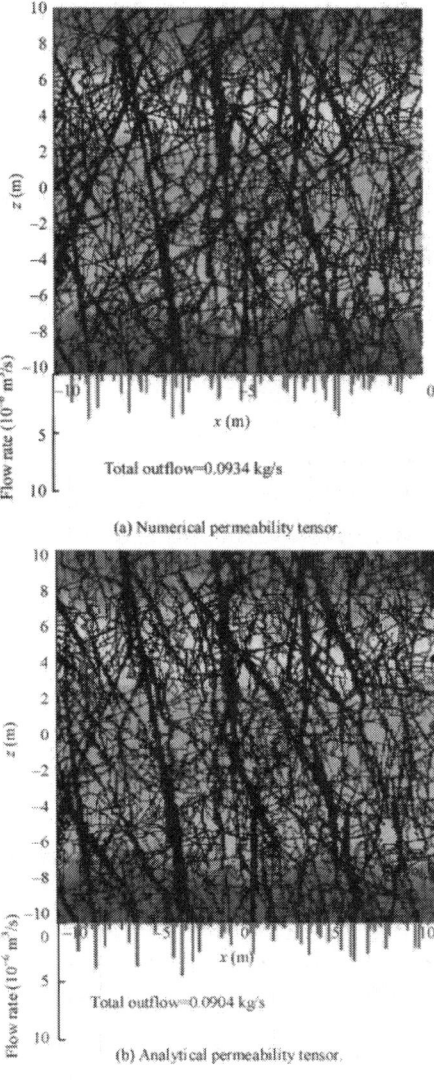

Figure 8. TOUGH2 steady state flow results under a vertical hydraulic gradient of 10 kPa/m for a 400 × 400 grid (160,000 grid-blocks and 0.05 m side-length) with comparison of results in which the permeability tensor was evaluated numerically (using an approach developed by Jackson et al., 2000) or analytically (using Oda's crack tensor theory due to Oda, 1986). The pressure goes from 0.3 MPa (red) to 0.1 MPa (blue) resulting in a pressure gradient of 10 kPa/m. The thickness of the black lines within the model domain are proportional to the mass flow rate and the distribution of flow is shown at the bottom in terms of volumetric rate (m³/s) flowing out of each 0.05 m sized grid element at the bottom of the model domain.

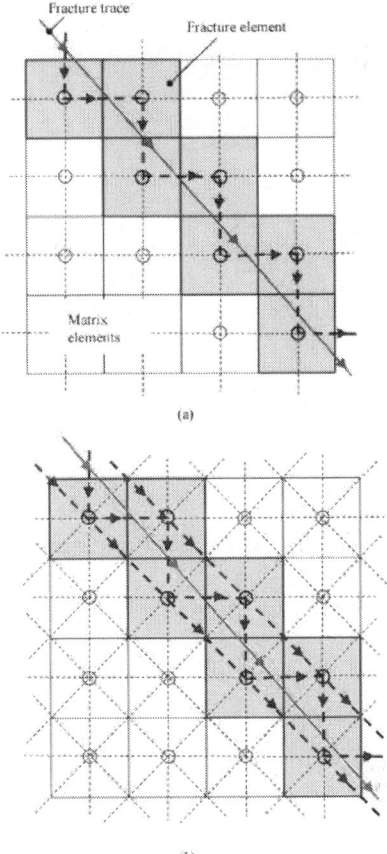

Figure 9. Schematics of FDM schemes in TOUGH2 applied in this study: (a) schematic representation showing how flow along a diagonal fracture trace would require longer flow distance along vertical and horizontal connections following a staircase path when using 5-point FDM grid, and (b) 9-point differencing that includes additional diagonal connections.

When element size increased, the total flow rate through the model stabilized to values in reasonable agreement with the NAPSAC DFN simulation results (Fig. 10). The flow rate was stabilized for element side lengths of 0.5 m or larger (i.e. when the model was divided into 40 × 40 elements or less). The 40 × 40 element case was chosen for a more detailed comparison with other DFN results, including fluid flow and transport changes under increasing load. In the 40 × 40 element case, the total flow rate through the

20 m × 20 m model is similar to the DFN, so the average solute transit time should be comparable. But as shown in Fig. 10, some of the heterogeneities have been smoothed out, which in theory could lead to a less dispersed (sharper) breakthrough.

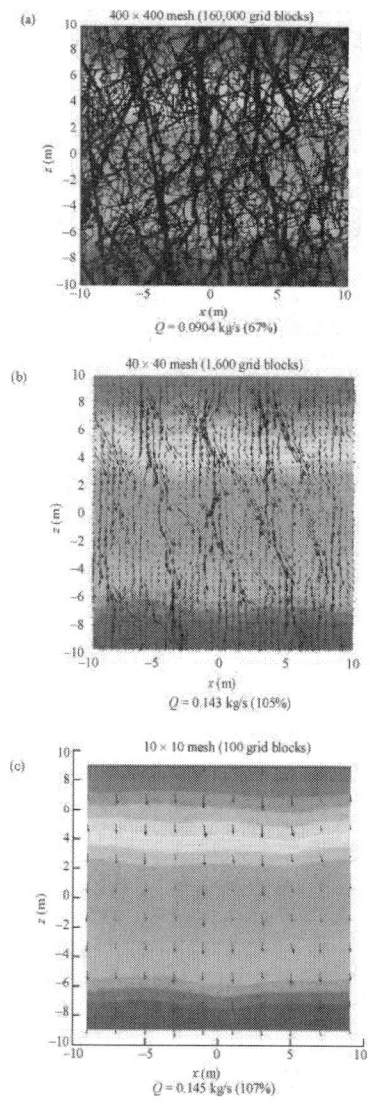

Figure 10. TOUGH2 flow distribution and total flow rates through the bottom boundary for different grid sizes and comparison to total mass flow rate to DFN model results in the case of a vertical hydraulic gradient of 10 kPa/m. The total TOUGH2 flow is given below each model case and compared in terms of percentage of the total NAPSCA DFN model results.

Fluid flow under increasing mechanical load

Fig. 11 presents the results of horizontal and vertical permeability calculated from the downstream boundary flow rate under horizontal and vertical gradients, respectively. The figure shows that, overall, permeability is high (in the order of 1×10^{-13} m²), and that under increasing mechanical load, the vertical permeability decreases more than the horizontal. This is reasonable, considering that most of the loading is horizontal and therefore tends to preferentially compress vertical fractures to a smaller aperture.

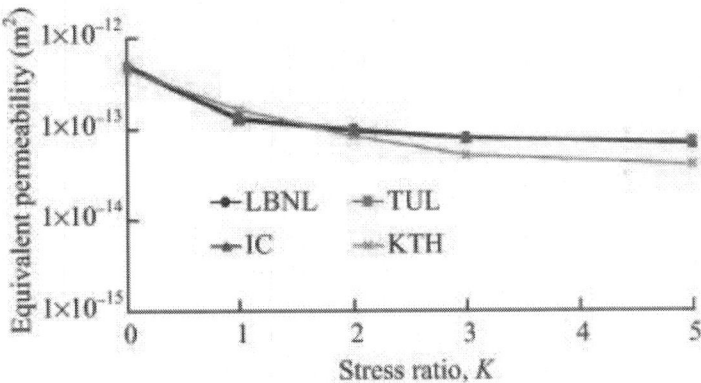

(a) Horizontal hydraulic gradient.

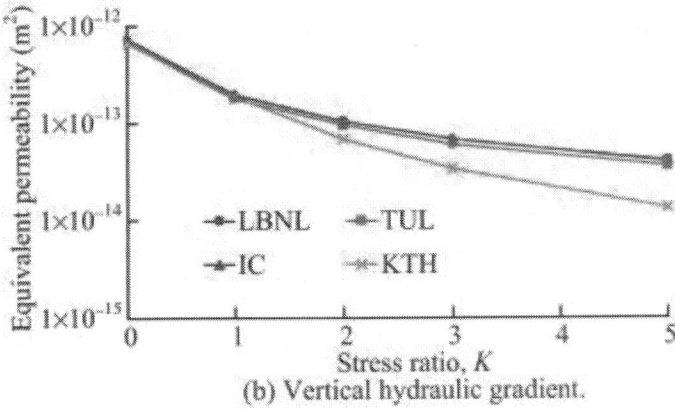

(b) Vertical hydraulic gradient.

Figure 11. Model simulation comparison of load dependent equivalent permeability calculated from the downstream boundary flow rate in the case of horizontal and vertical hydraulic gradients.

Moreover, TOUGH-FLAC modeling results for the 40×40 element model is in close agreement with alternative DFN models, even with increasing stress ratios. The TUL results shown in Fig. 11 correspond to TUL_2 in Zhao et al. (2013), in which the rock mass shear stiffness was chosen to be similar to that of IC and LBNL. The results by the KTH team, using the UDEC DFN model simulation, show a somewhat larger decrease in permeability than other teams. This might be caused by a difference in the normal closure model (the KTH team used a tri-linear approximation of the hyperbolic closure model), or it might result from the UDEC model based on the distinct element method (DEM), the only one of the four models that allows for block rotations, which might cause additional localized fracture closure at block corners (Zhao et al., 2013).

The results also show that shear dilation has a minor impact on permeability in this case. This might be a surprising result, considering that the highest horizontal-to-vertical stress ratio of 5 is quite extreme. In fact, shear failure along optimally oriented fractures (fractures dipping about 30°) started to occur at a stress ratio of 3, and a large number of fractures had failed in shear at a stress ratio of 5. However, while many fractures are in a shear failure mode, most of them are smaller fractures confined within the rock mass, which prevents them from sliding sufficiently to cause significant shear dilation. Moreover, small fractures that might be sheared may connect to other fractures of different orientations that are not sheared, and no continuous path of shear-dilated fractures forms.

With a simple calculation using Eq. (16), we can find an explanation for the lack of shear induced permeability in this case. For example, at the highest stress ratio, i.e. 5 MPa of vertical stress and 25 MPa of horizontal stress, the maximum shear stress upon an optimally oriented fracture will be

$$\tau_m = \frac{1}{2}(\sigma_1 - \sigma_3) = \frac{1}{2}(25 - 5) = 10\,\text{MPa} \tag{18}$$

Shear failure for such a case can be determined using the Mohr–Coulomb criterion, written in the form of:

$$\sigma_1 = \frac{2c\cos\varphi}{1 - \sin\varphi} + \frac{1 + \sin\varphi}{1 - \sin\varphi}\sigma_3 \tag{19}$$

which for $c = 0$ and $\varphi = 24.9°$ yields $\sigma_1 = 2.45\sigma_3$. Consequently, for a vertical stress of 5 MPa, failure would occur at a horizontal stress of $2.45 \times 5 = 12.25$ MPa, and in that instance, the maximum shear stress would be $\tau_m = 3.6$ MPa. This means that a shear-stress increment of $\Delta\tau = 10 - 3.6 = 6.4$ MPa can occur after shear failure.

For a mean fracture length of 0.92 m, we previously estimated $K_s^{rock} = 67.8\,GPa/m$, which, according to Eq. (16), would result in a dilation of $\Delta b_{dil} = (\Delta\tau/K_s^{rock}) \times \tan\varphi_d = (6.4 \times 10^6/67.8 \times 10^9)\tan 5° \approx 8.2 \times 10^{-6}\,m$, i.e. 8.2 µm. According to Fig. 2, a 0.92 m long fracture has an initial aperture of about 30 µm, which means that transmissivity through such a fracture might increase according to the cubic law by about $[(30 + 8.2)/30]^3 \approx 2.1$, i.e. a factor of 2.1.

For a 20 m fracture, the initial aperture is already close to the maximum allowable aperture of 200 µm, so no significant increase can occur. This means that the permeability of the most open and permeable large scale fractures will not increase very much even optimally oriented for shear. As a result, the overall permeability change by shear dilation is expected to be quite small in this case.

Fracture-dominant advective transport

Under the relatively high gradient of 10 kPa/m, the solute transport is dominated by advection within the fracture system. This is illustrated in Fig. 12, which shows the results of the total solute transport through the fracture system, with a negligible amount of solute residing in the matrix. Thus, under this high hydraulic gradient, matrix diffusion and fracture–matrix interactions are negligible.

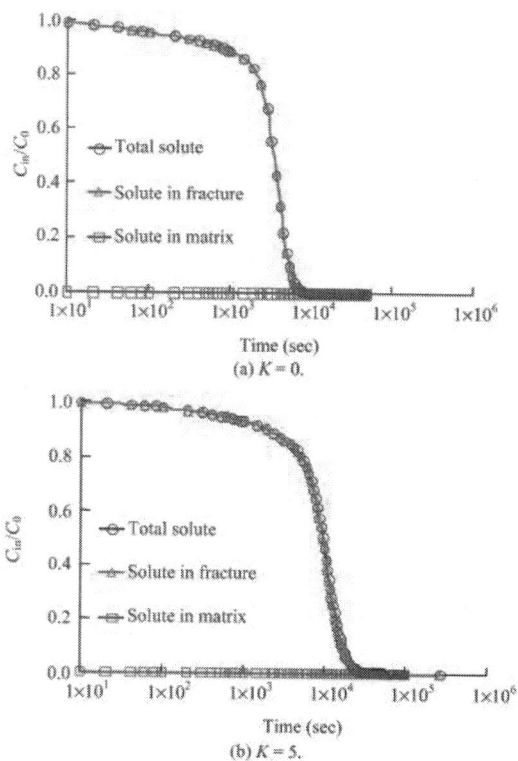

Figure 12. Solute concentration variation in fractures and rock matrix under a horizontal hydraulic gradient of 10 kPa/m.

Fig. 13 presents breakthrough curves for the three outlet boundaries with comparison to the results of the three alternative DFN models. For $K = 0$, which corresponds to the initial unstressed case, there is excellent agreement in the breakthrough curves between the different teams. At increasing stress ratios, some differences can be observed. In general, the results of our multicontinuum and solute transport approach are very similar to those of the TUL, even under increasing mechanical load. The results of KTH and IC indicate a relatively slower breakthrough under increasing load. It is not clear what causes this systematic difference in the breakthrough (considering that the permeability and total flow through the model domain are very similar for the different models). It appears that the two teams (KTH and IC) that used particle tracking obtained a delayed breakthrough compared to the two teams (LBNL and TUL) that used standard solute-transport simulation.

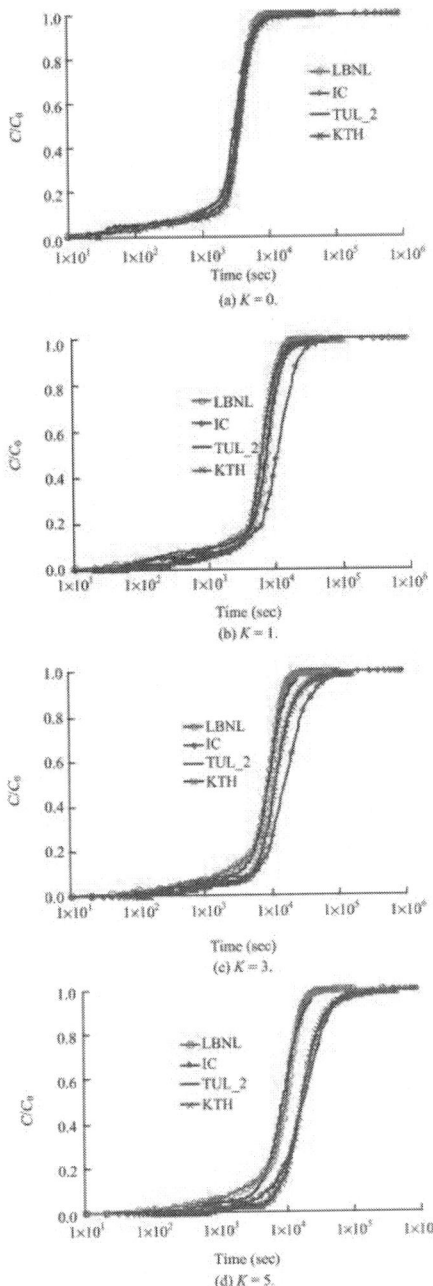

Figure 13. Comparison of breakthrough curves for tracers exiting from all the three outlets for stress ratios $K = 0$, 1, 3, and 5, under a horizontal hydraulic gradient of 10 kPa/m. DFN results (IC, TUL and KTH) used for comparison were extracted from Zhao et al. (2013).

Fig. 14 presents the results for breakthrough curves of different model discretizations. The figure shows that the breakthrough becomes sharper as element sizes decrease. This might be counterintuitive, because for smaller grid sizes we should be able to resolve a heterogeneous flow pattern more precisely, and thereby the breakthrough should be more dispersed. The results in Fig. 14 show that the breakthrough curves for the 100 × 100 grid are somewhat delayed as a result of the above-discussed reduced total flow rate in the case of finer mesh. Moreover, as shown in Fig. 15, for very large element sizes, we see some additional fraction of injected solute exit by diagonal flow through lateral boundaries. Specifically, Fig. 15b shows a relatively larger fraction of the solute exit through the top and bottom boundaries, which are lateral boundaries relative to the inlet, and consequently a relatively smaller fraction exit through the left boundary, which in this case is the downstream outlet boundary. However, one critical issue with large grid sizes, such as in the cases of 10 × 10 and 4 × 4 grids, is numerical dispersion, which in fact is the main cause of the observed more disperse breakthrough curves.

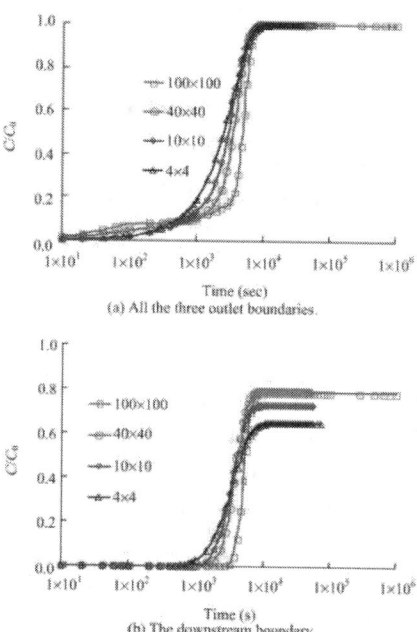

Figure 14. Effects of grid size on the breakthrough curves of tracers exiting from all the three outlet boundaries and the downstream boundary under a horizontal hydraulic gradient of 10 kPa/m and no mechanical load ($K = 0$).

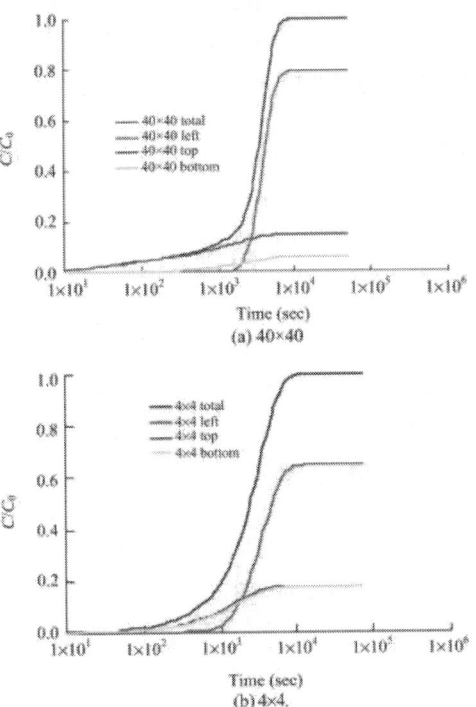

Figure 15. Breakthrough curves of tracers exiting through various boundaries under a horizontal hydraulic gradient of 10 kPa/m and no mechanical load ($K = 0$) in the cases of 40×40 and 4×4 element discretization.

We investigated the effect of numerical dispersion by simulations in which we distinguish between the regular FDM numerical grid and a property grid in which the material properties, including permeability and elastic tensors, are homogenized and defined. Fig. 16 presents two such cases. In Fig. 16a, we present the results for a 10×10 property grid, whereas the numerical grid is either 10×10 or 100×100. Likewise, in Fig. 16b, we present the results for a 4×4 property mesh, but with 4×4 or 40×40 numerical grid. The results in Fig. 16 show that a much sharper breakthrough is obtained for a finer numerical grid resolution, signifying the profound effect of numerical dispersion in the transport calculation. Remedies to cope with numerical dispersion exist, such as total-variation-diminishing or flux-limiter schemes, which have also been implemented in special versions of the TOUGH2 code (Wu and Forsyth, 2006). Without such remedies, another possibility would be to use different grid resolutions for the numerical analysis and evaluation of equivalent properties. That is, to use a fine numerical grid that can resolve sharp fronts of solute transport, and to use a larger grid (a property grid) to evaluate continuum properties.

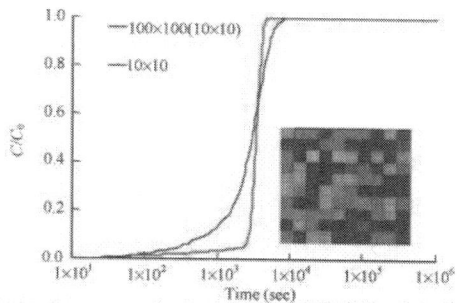

(a) 10 ×10 property mesh with 10 ×10 or 100×100 FDM grids blocks and 3 outlet boundaries.

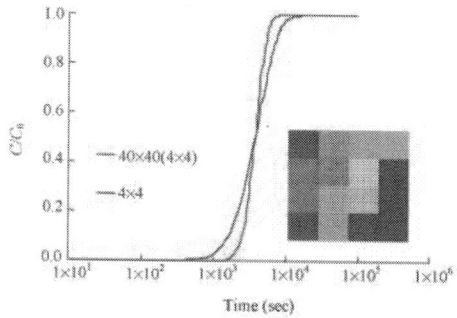

(b) 4 ×4 property mesh with 4 ×4 or 40×400 FDM grids blocks and one outlet boundary.

Figure 16. Investigation of the effects of numerical dispersion by comparing breakthrough under a horizontal hydraulic gradient of 10 kPa/m and no mechanical load ($K = 0$) using a property mesh and different numerical grid resolutions: 10 × 10 property mesh with 10 × 10 or 100 × 100 FDM grids blocks and 3 outlet boundaries, and 4 × 4 property mesh with 4 × 4 or 40 × 40 FDM grid blocks and one outlet boundary.

Fracture–matrix diffusion retarded transport

Fig. 17 and Fig. 18 present simulation results for the case of a much lower hydraulic gradient of 10 Pa/m. In this case, the transit time through the network increases, and matrix diffusion becomes much more significant. Fig. 18 shows that the importance of matrix diffusion increases with load, because fracture permeability decreases and transit time increases. For the high load case (Fig. 18b), at one point of time as much as 60% of the injected solute resides within the matrix, causing a significant additional delay in the breakthrough. Note that in this case, no advective transport takes place between fracture and matrix, because the pressure is steady and uniform at all times (thus, there is no pressure gradient between fractures and matrix).

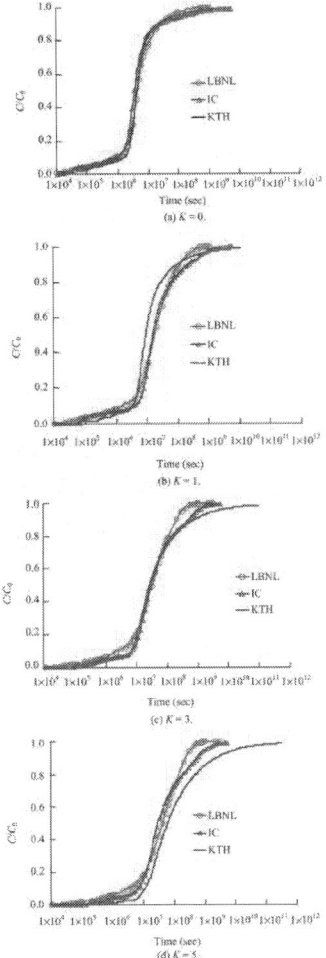

Figure 17. Comparison of breakthrough curves for tracers exiting from all the three outlet boundaries for stress ratios K = 0, 1, 3, and 5, under a horizontal hydraulic gradient of 10 Pa/m. DFN results (IC and KTH) used for comparison were extracted from Zhao et al. (2013).

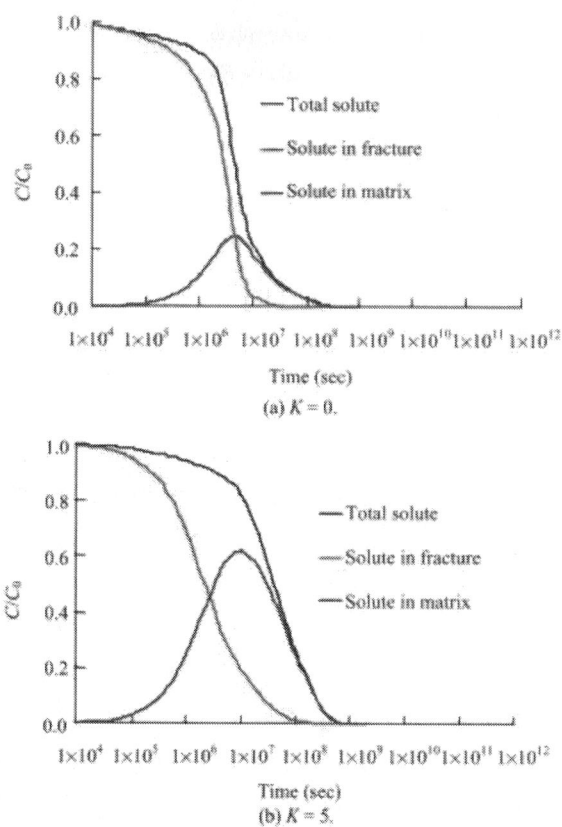

Figure 18. Solute concentration variation in fractures and rock matrix when $K = 0$ and 5 under a horizontal hydraulic gradient of 10 Pa/m.

The comparison of breakthrough curves between our multicontinuum model results and two DFN modeling results, in Fig. 17, shows good agreement in overall transport behavior, in particular at $K = 0$ (Fig. 17a). Note that in the DFN model simulations, the rock matrix is not discretized, but matrix diffusion is modeled using a one-dimensional diffusion model that results in an extra retarded travel time. The good agreement in the matrix-diffusion results between the multicontinuum and DFN approaches is encouraging, because the low gradient case of strong fracture–matrix interaction behavior is more realistic from a repository-performance perspective. A dependency on numerical grid size was also observed for the low gradient case, affecting the advective transport in the fracture system, whereas grid size is not expected to directly impact the calculated matrix diffusion.

CONCLUDING REMARKS

In this paper, we presented and applied a linked multicontinuum and crack tensor approach for modeling of coupled geomechanics, fluid flow, and solute transport in fractured rock. We used the crack tensor theory to calculate effective block-scale properties, including permeability and elastic tensors, as well as multicontinuum properties relevant to matrix diffusion. In the modeling, we considered stress dependent properties through stress-induced changes in fracture apertures that were used to update permeability and elastic tensors as a function of mechanical load. We evaluated the effectiveness and accuracy of our linked multicontinuum and crack tensor approach by comparing our modeling results with those of three independent DFN models.

In two of the three alternative DFN models, solute transport was simulated using particle tracking, i.e. a very different method from the standard solute transport used in our multicontinuum modeling. We compared the results for flow and solute transport through a 20 m × 20 m model domain of fractured rock — with detailed comparison of total flow rate, its distribution, and breakthrough curves. In our modeling, we divided the 20 m × 20 m model domain into regular blocks, or continuum elements for the numerical simulations. We selected a model discretization of 40 × 40 elements (having a side length of 0.5 m), which resulted in flow rates equivalent to that of the DFN models. Using such a model discretization, our results were in reasonably good agreement with the alternative DFN models, for advective dominated transport (under high hydraulic gradient) as well as for matrix diffusion retarded transport (under low hydraulic gradient). We think these are encouraging results, especially since our modeling approach could be readily applied for studying large-scale coupled processes in three dimensions. Of course, we must also consider that in using the large block-scale homogenized properties, we miss out on some of the heterogeneities, leading to a less dispersed transport. One potential remedy is the possibility of using the crack tensor to calculate the dispersivity of each element and adding such capability to the solute transport modeling.

However, we also found that numerical dispersion has a pronounced impact on the flow when using larger grid spacing. Thus, it appears that the best way to cope with these issues is to use a proper mesh that is large enough to achieve representative continuum properties and a numerical grid that is sufficiently fine to minimize numerical dispersion.

Finally, using the crack tensor and TOUGH2 finite volume scheme, we encountered some difficulties when element sizes are so small that only one or a few fractures would intersect an element. In such a case, we are practically modeling the fractures discretely, albeit approximately, and this is an area where improvements are currently being pursued.

ACKNOWLEDGMENTS

The work described in this paper was conducted within the context of the international DECOVALEX-2011 Project. Financial support for Berkeley Laboratory authors was provided by the UK Nuclear Decommissioning Authority (NDA) to the Lawrence Berkeley National Laboratory through the National Energy Technology Laboratory, under the U.S. Department of Energy Contract No. DE-AC02-05CH11231. The work of Colin Leung was funded by the UK Engineering and Physical Sciences Research Council (EPSRC), the UK Nuclear Decommissioning Authority (NDA), and Serco, through CASE studentship award 08002638; particular thanks go to Cherry Tweed and Simon Norris of the NDA for supporting this work. Editorial review by Dan Hawkes at the Berkeley Laboratory is greatly appreciated. The views expressed in the paper are however those of the authors and are not necessarily those of the Funding Organizations.

REFERENCES

1. Baghbanan A, Jing L. Hydraulic properties of fractured rock masses with correlated fracture length and aperture. International Journal of Rock Mechanics and Mining Sciences 2007;44(5):704–19.

2. Baghbanan A, Jing L. Stress effects on permeability in a fractured rock mass with correlated fracture length and aperture. International Journal of Rock Mechanics and Mining Sciences 2008;45(8):1320–34

3. Bai M. On equivalence of dual-porosity poroelastic parameters. Journal of Geophysical Research 1999;104(B5):10461–6.

4. Bandis SC, Lumsden AC, Barton NR. Fundamentals of rock joint deformation. International Journal of Rock Mechanics and Mining Sciences and Geomechanics Abstracts 1983;20(6):249–68.

5. Barenblatt GI, Zheltov IP, Kochina IN. Basic concepts in the theory of seepage of homogeneous liquids in fissured rocks. Journal of Applied Mathematics 1960;24(5):1286–303.

6. Barton NR, Bandis S, Bakhtar K. Strength, deformation and conductivity coupling of rock joints. International Journal of Rock Mechanics and Mining Sciences and Geomechanics Abstracts 1985;22(3):121–40.

7. Berryman JG, Wang HF. The elastic coefficients of double-porosity models for fluid transport in jointed rock. Journal of Geophysical Research 1995;100(B12):24611–27.

8. Berryman JG. Extension of poroelastic analysis to double-porosity materials: new technique in microgeomechanics. Journal of Engineering Mechanics: ASCE 2002;128(8):840–7.

9. Biot MA. A general theory of three-dimensional consolidation. Journal of Applied Physics 1941;12(2):155–64.

10. Blum P, Mackay R, Riley MS. Stochastic simulations of regional scale advective transport in fractured rock masses using block upscaled hydro-mechanical rock property data. Journal of Hydrology 2009;369(3–4):318–25.

11. Chen HY, Teufel LW. Coupling fluid flow and geomechanics in dual-porosity modeling of naturally fractured reservoirs. In: SPE annual technical conference and exhibition, 1997. San Antonio, London: Society of Petroleum Engineers; 1997. p. 419–33.

12. Dieterich JH. Earthquake nucleation on faults with rate- and state-dependent friction. Tectonophysics 1992;211:115–34.

13. Guvanasen V, Chan T. Upscaling the THM properties of a fractured rock mass using a modified crack tensor theory. In: Stephansson O, Hudson JA, Jing L, editors. Coupled thermo-hydro-mechanical-chemical processes in geo-systems. Oxford: Elsevier; 2004. p. 251–6.

14. Itasca Consulting Group. FLAC3D, fast Lagrangian analysis of continua in 3 dimensions (ver. 4.0). Minneapolis, Minnesota: Itasca Consulting Group; 2009. p. 438.

15. Jackson CP, Hoch AR, Todman S. Self-consistency of a heterogeneous continuum porous medium representation of a fractured medium. Water Resources Research 2000;36(1):189–202.

16. Jing L, Hudson JA. Numerical methods in rock mechanics. International Journal of Rock Mechanics and Mining Sciences 2002;39(4):409–27.

17. Kim J, Sonnenthal E, Rutqvist J. Formulation and sequential numerical algorithms of coupled fluid/heat flow and geomechanics for multiple porosity materials. International Journal of Numerical Methods in Engineering 2012;92(5):425–56.

18. Kobayashi A, Fujita T, Chijimatsu M. Continuous approach for coupled mechanical and hydraulic behavior of a fractured rock mass during hypothetical shaft sinking at Sellafield, UK. International Journal Rock Mechanics and Mining Sciences 2001;38(1):45–57.

19. Liu HH, Rutqvist J. Coupled hydro-mechanical processes associated with multiphase flow in a dual-continuum system: formulations and a sensitivity study. Rock Mechanics and Rock Engineering 2012., http://dx.doi.org/10.1007/s00603-012-0313-3. Min KB, Rutqvist J, Tsang CF, Jing L. Stress-dependent permeability of fracture rock masses: a numerical study. International Journal of Rock Mechanics and Mining Sciences 2004;41(7):1191–210.

20. Min KB, Rutqvist J, Tsang CF, Jing L. Thermally induced mechanical and permeability changes around a nuclear waste repository—a far-field study based on equivalent properties determined by a discrete approach. International Journal of Rock Mechanics and Mining Sciences 2005;42(5–6):765–80.

21. Neretnieks I. Diffusion in the rock matrix: an important factor in radionuclide retardation. Journal of Geophysical Research 1980;85(B8):4379–97.

22. Noorishad J, Ayatollahi MS, Witherspoon PA. A finite element method for coupled stress and fluid flow analysis of fractured rocks.

International Journal of Rock Mechanics Mining Sciences and Geomechanics Abstracts 1982;19(4):185–93.

23. Oda M. An equivalent continuum model for coupled stress and fluid-flow analysis in jointed rock masses. Water Resources Research 1986;22(13):1845–56.

24. Öhman J, Niemi A. Upscaling of fracture hydraulics by means of an oriented correlated stochastic continuum model. Water Resources Research 2003;39(10):1277–88. Pruess K, Narasimhan TN. A practical method for modeling fluid and heat-flow in fractured porous-media. SPE Journal 1985;25(1):14–26.

25. Pruess K, Oldenburg C, Moridis G. TOUGH2 user's guide (ver. 2.0). Berkeley, USA: Lawrence Berkeley National Laboratory; 1999. p. 210.

26. Pruess K. Grid orientation and capillary pressure effects in the simulation of waster injection into depleted vapors zones. Geothermics 1991;20(5–6):257–77.

27. Rutqvist J, Barr D, Birkholzer JT, Fujisaki K, Kolditz O, Liu QS, et al. A comparative simulation study of coupled THM processes and their effect on fractured rock permeability around nuclear waste repositories. Environmental Geology 2009a;57(6):1347–60.

28. Rutqvist J, Freifeld B, Min KB, Elsworth D, Tsang Y. Analysis of thermally induced changes in fractured rock permeability during eight years of heating and cooling at the Yucca Mountain Drift Scale Test. International Journal of Rock Mechanics and Mining Sciences 2008;45(8):1373–89.

29. Rutqvist J, Wu YS, Tsang CF, Bodvarsson G. A modeling approach for analysis of coupled multiphase fluid flow, heat transfer, and deformation in fractured porous rock. International Journal of Rock Mechanics and Mining Sciences 2002;39(4):429–42.

30. Rutqvist J, Tsang CF. Analysis of thermal-hydrologic-mechanical behavior near an emplacement drift at Yucca Mountain. Journal of Contaminant Hydrology 2003;62–63:637–52.

31. Rutqvist J, Stephansson O. The role of hydromechanical coupling in fractured rock engineering. Hydrogeology Journal 2003;11(1):7–40.

32. Rutqvist J, Barr D, Datta R, Gens A, Millard M, Olivella S, et al. Coupled thermal-hydrological-mechanical analysis of the Yucca Mountain Drift Scale Test—comparison of field results to predictions of four different models. International Journal of Rock Mechanics and Mining Sciences 2005;42(5–6):680–97.

33. Rutqvist J, Bäckström A, Chijimatsu M, Feng XT, Pan PZ, Hudson J, et al. A multiplecode simulation study of the long-term EDZ evolution of geological nuclear waste repositories. Environmental Geology 2009b;57(6):1313–24.

34. Rutqvist J. Status of the TOUGH-FLAC simulator and recent applications related to coupled fluid flow and crustal deformations. Computers and Geosciences 2011;37(6):739–50.

35. Rutqvist J. The geomechanics of CO2 storage in deep sedimentary formations. Geotechnical and Geological Engineering 2012;30(3):525–51.

36. Stietel A, Millard A, Treille E, Vuillod E, Thoravel A, Ababou R. Continuum representation of coupled hydromechanical processes of fractured media: homogenisation and parameter identification. In: Stephansson O, Jing L, Tsang CF, editors. Developments in geotechnical engineering: coupled thermo-hydro-mechanical processes of fractured media. Amsterdam: Elsevier Science B.V.; 1996. p. 135–64.

37. Walsh JB. Effects of pore pressure and confining pressure on fracture permeability. International Journal of Rock Mechanics and Mining Sciences and Geomechanics Abstracts 1981;18(5):429–35.

38. Warren JE, Root PJ. The behavior of naturally fractured reservoirs. SPE Journal 1963;3(3):245–55.

39. Wu YS, Forsyth PA. Efficient schemes for reducing numerical dispersion in modeling multiphase transport through porous and fractured media. In: Proceedings of the TOUGH symposium 2006.

40. Berkeley, USA: Lawrence Berkeley National Laboratory; 2006. p. 1–27. Zhang X, Sanderson DJ. Scale up of two-dimensional conductivity tensor for heterogenous fracture networks. Engineering Geology 1999;53(1):83–99.

41. Zhao Z, Rutqvist J, Leung C, Hokr M, Neretnieks I, Hoch A, et al. Stress effects on solute transport in fractured rocks: a comparison study. Journal of Rock Mechanics and Geotechnical Engineering 2013;5(2); in press.

CITATION

Jonny Rutqvist, Colin Leung, Andrew Hoch, Yuan Wang, Zhen Wang, Linked multicontinuum and crack tensor approach for modeling of coupled geomechanics, fluid flow and transport in fractured rock, Journal of Rock Mechanics and Geotechnical Engineering, Volume 5, Issue 1, February 2013, Pages 18-31, ISSN 1674-7755, http://dx.doi.org/10.1016/j.jrmge.2012.08.001.

CHAPTER 6

Numerical Modeling for the Coupled Thermo-Mechanical Processes and Spalling Phenomena in Äspö Pillar Stability Experiment (Apse) ☆

T. Koyama[1], M. Chijimatsu[2], H. Shimizu[3], S. Nakama[4], T. Fujita[5], A. Kobayashi[e], Y. Ohnishi[6]

[1] Department of Urban Management, Kyoto University, Kyoto, Japan

[2] Hazama Corporation, Tokyo, Japan

[3] Institute of Fluid Science, Tohoku University, Sendai, Japan

[4] Japan Atomic Energy Agency (JAEA), Tokai, Japan

[5] Faculty of Environmental and Urban Engineering, Kansai University, Osaka, Japan

[6] Kyoto University, Kyoto, Japan

ABSTRACT

In this paper, the coupled thermo-mechanical (TM) processes in the Äspö Pillar Stability Experiment (APSE) carried out by the Swedish Nuclear Fuel and Waste Management Company (SKB) were simulated using both continuum and discontinuum based numerical methods. Two-dimensional (2D) and three-dimensional

(3D) finite element method (FEM) and 2D distinct element method (DEM) with particles were used. The main objective for the large scale in situ experiment is to investigate the yielding strength of crystalline rock and the formation of the excavation disturbed/damaged zone (EDZ) during excavation of two boreholes, pressurizing of one of the boreholes and heating. For the DEM simulations, the heat flow algorithm was newly introduced into the original code. The calculated stress, displacement and temperature distributions were compared with the ones obtained from in situ measurements and FEM simulations. A parametric study for initial microcracks was also performed to reproduce the spalling phenomena observed in the APSE.

INTRODUCTION

The DECOVALEX-2011 is an international cooperative research project on mathematical and numerical models of coupled thermo-hydro-mechanical-chemical (THMC) processes for safety analysis of radioactive waste repositories. The DECOVALEX project evolved from 1992 and DECOVALEX-2011 is the fifth stage (after DECOVALEX I, II, III and THMC). Task B in DECOVALEX-2011 was defined for simulating the thermo-elastic behavior of the Äspö Pillar Stability Experiment (APSE) performed at a depth of 450 m in the Äspö Hard Rock Laboratory of the Swedish Nuclear Fuel and Waste Management Company (SKB) (Andersson, 2007). The main objective of the APSE is to investigate the yielding strength of crystalline rock, the formation and growth of the excavation disturbed/damaged zone (EDZ) and spalling phenomena during excavation and heating processes. The experimental layout consists of a tunnel with arched roof and floor (7.5 m high and 5 m wide) as well as two 1.75 m diameter (6.5 m and 6.3 m deep, respectively) boreholes separated by a 1.0 m thick pillar of Äspö diorite (Fig. 1). Because of the relatively low in situ stresses compared with the strength of intact rock, specially designed excavation- and thermal-induced stresses were introduced to ensure stress magnitudes sufficient to induce spalling of the rock mass. Hence, both elastic and non-elastic rock mass responses could be captured as the magnitudes of excavation- and thermal-induced stresses were gradually increased (Andersson, 2007).

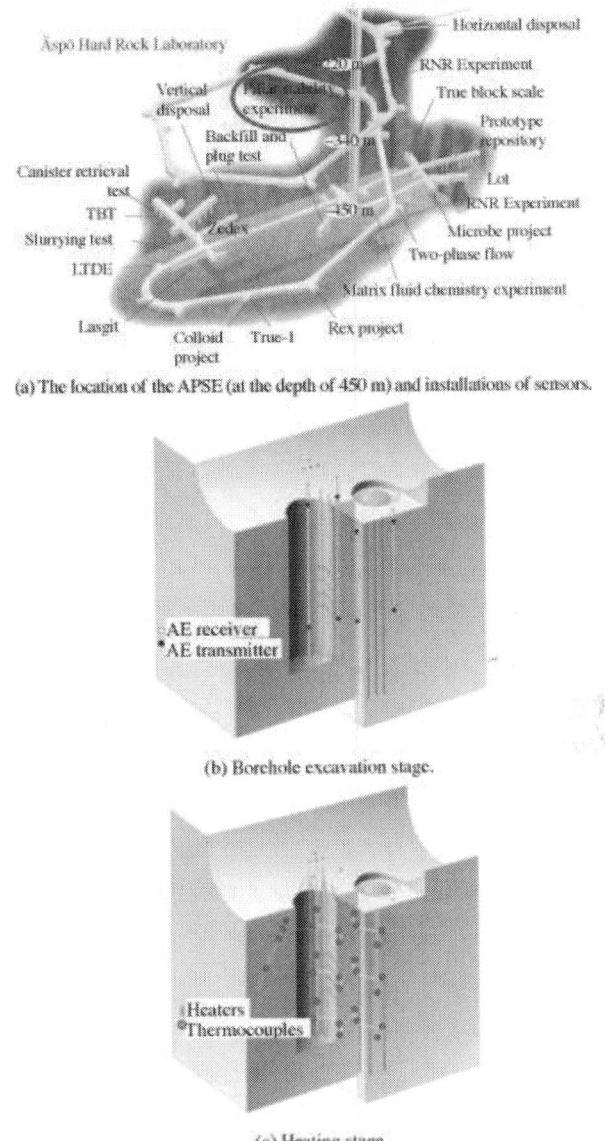

(a) The location of the APSE (at the depth of 450 m) and installations of sensors.

(b) Borehole excavation stage.

(c) Heating stage.

Figure 1. APSE performed at the depth of 450 m in the Äspö Hard Rock Laboratory, Sweden (Andersson, 2007).

The in situ experiment was carried out in the following stages: (1) excavation of the test tunnel, (2) excavation of the first vertical borehole of 1.75 m in diameter and 6.5 m in depth, (3) installation of the confining system and application of confining pressure, (4) excavation of the second borehole of 1.75 m in diameter and 6.3 m in depth to form the 1.0 m thick pillar between the boreholes, and (5) increase of the temperature by several electric heaters installed in several boreholes with small diameters. The detail description of the experiment can also be seen in Andersson (2007), Andersson and Martin (2009), and Andersson et al. (2009).

In this paper, the coupled thermo-mechanical (TM) processes in the APSE were simulated using both continuum and discontinuum based numerical methods; two-dimensional (2D) and three-dimensional (3D) finite element method (FEM) codes called THAMES (Kobayashi and Ohnishi, 1986 and Ohnishi et al., 1987) with damage mechanics and 2D distinct element method (DEM) with particles (Potyondy and Cundall, 2004 and Shimizu et al., 2010) were used. The mechanical parameters for these simulations such as damage parameters (for FEM simulations) and microscopic parameters (for DEM simulations) were calibrated using laboratory uniaxial compressive tests for rock core sample (Äspö diorite) obtained from the site (Staub et al., 2004). The FEM can treat coupled thermo-hydro-mechanical (THM) processes in saturated–unsaturated porous media. On the other hand, the DEM can treat the rock heterogeneity easily and investigate the failure mechanism in detail, including crack initiation and propagation. The thermal flow algorithms were newly introduced into the original DEM code to investigate the failure mechanism during heating (Shimizu et al., 2011). The simulation results such as stress distribution, displacements as well as temperature distribution and their time evolution during excavation and heating phases by 2D/3D FEM and 2D DEM were compared with in situ measurements.

CONTINUUM BASED APPROACH, 2D/3D FEM SIMULATIONS

Governing equations
In this study, to simulate the TM processes in the APSE, the FEM code THAMES (Kobayashi and Ohnishi, 1986 and Ohnishi et al.,

1987) was used. THAMES can treat fully coupled THM behaviors of a saturated–unsaturated medium. The mathematical formulation for THAMES is based on Biot's theory with the Duhamel–Neuman's form of Hooke's law and energy balance equations. The principal unknowns are total pressure, displacements and temperature. THAMES has been extended to consider the thermo-osmosis in the buffer materials such as water movement due to thermal gradient as well as swelling phenomena and validated with the data obtained from laboratory tests (Chijimatsu et al., 1998), the engineered scale tests (Chijimatsu et al., 2000a) as well as the in situ experiments (Chijimatsu et al., 2000b).

Conservation law
Conservation law of momentum satisfies the static stress equilibrium for macroscopic total stresses as follows:

$$\Delta \sigma_{ij,j} = 0 \tag{1}$$

where $\Delta \sigma_{ij}$ is the tensor of total stress increment.

The mass conservation for fluid and the heat energy conservation can be described as

$$\frac{\partial(\theta \rho_w)}{\partial t} + q_{i,i} + Q = 0 \tag{2}$$

$$\frac{\partial[(\rho c)_m T]}{\partial t} + q^h_{i,i} + Q^h = 0 \tag{4}$$

where θ is the volumetric water content; ρ_w is the water density; t is the time; q_i is the groundwater flux vector; Q is the sink/source; $(\rho c)_m$ is the specific heat of mixture of gas, liquid and solid; T is the temperature; q^h_i is the heat flux vector; and Q^h is the heat sink/source.

Constitutive equations
Assuming the rock is elastic material, the increment of total stress can be written as follows:

$$\Delta \sigma_{ij} = \frac{1}{2} C_{ijkl} (\Delta u_{k,l} + \Delta u_{l,k}) + S_r \Delta p \delta_{ij} - \beta \Delta T \delta_{ij} \tag{5}$$

$$\beta=(3\lambda+2\mu)\alpha_s \qquad (6)$$

where C_{ijkl} is the elastic modulus tensor, u_i is the displacement, δ_{ij} is the Kronecker's delta, λ and μ are the Lame's constants, α_s is the thermal expansion coefficient, p is the water pressure, and S_r is the degree of saturation.

Groundwater flux (q_i in Eq. (2)) can be calculated using the following equation:

$$q_i = -\frac{\rho_w^2 g k_r K_{ij}}{\mu_w} h_{,j} - \rho_w (D_T)_{ij} T_{,j} \qquad (6)$$

where K is the intrinsic permeability; μ_w is the viscosity of water; g is the gravitational acceleration; $(DT)_{ij}$ is the thermal flow diffusivity; k_r is the relative permeability; h is the hydraulic head; $h_{,j}$ and $T_{,j}$ are the hydraulic and temperature gradients in space, respectively.

The first and second terms at the right side of Eq. (16) represent Darcy flow and thermo-osmosis, respectively.

Heat flux, q_i^h (the second term at the left side of Eq. (3)) can be divided into two terms, the advective term, q^a, and the dispersive term, q^d, as

$$q_i^h = q_i^a + q_i^d \qquad (7)$$

$$q_i^a = S_r q_i T \qquad (8)$$

$$q_i^d = -\lambda_m T_{,i} \qquad (9)$$

where λ_m is the thermal conductivity of rock mass, and $T_{,i}$ is the temperature gradient in space.

According to the damage mechanics theory, the change in mechanical behavior due to the growth of damage (microcracks) in material (damage expansion model) is considered and can be expressed using damage variable D (Lemaitre, 1992). The total strain ε_{ij} is assumed to be divided into the elastic strain, ε_{ij}^e, and the isotropic expansive strain, ε_{kk}^v, as

$$\varepsilon_{ij} = \varepsilon_{ij}^c + \frac{\varepsilon_{kk}^v}{3}\delta_{ij}$$

(10)

where expansive strains are negative by convention.

For the isotropic damage evolution, the relation between total stress, σ_{ij}, and strain, ε_{ij}, is expressed as

$$\sigma_{ij} = (1 - D)\left[(\lambda\varepsilon_{kk}\delta_{ij} + 2\mu\varepsilon_{ij}) - \frac{1}{3}\varepsilon_{kk}^v(3\lambda + 2\mu)\delta_{ij}\right]$$

(11)

where D is the damage variable.

The expansive strain ε_{kk}^v during damage progress will be equal to $3\varepsilon_{11}^v$ because ε_{ij}^v is isotropic. Hence, the equivalent damage conjugate force, Y_{eq}, can be described using D as

$$Y_{eq} = K_d D n_d + B_0$$

(12)

where K_d and n_d depend on damage evolution, and B_0 is the initial damage potential. Y_{eq} can also be expressed as a function of ε_{kk}^v as follows:

$$Y_{eq} = K_v\left(-\frac{1}{3}\varepsilon_{kk}^v\right)^{n_v}$$

(13)

where K_v is the gradient, and n_v is the scaling factor of the volumetric strain due to damage progress.

In this paper, K_d, n_d, B_0, K_v and n_v are so-called damage parameters. When Y_{eq} exceeds $B + B_0$, the damage variables increase (Murakami and Kamiya, 1997):

$$F(Y_{ij}, B) = Y_{eq} - (B_0 + B) = 0$$

(14)

where B is the difference between present damage potential Y_{eq} and previous damage potential B_0, and the equivalent damage conjugate force Y_{eq} is

$$Y_{eq} = \sqrt{\frac{1}{2}Y_{ij}Y_{ij}}$$

(15)

Hence, the damage conjugate force Y_{ij} can be given as

$$Y_{ij} = -\frac{1}{2}(\lambda \varepsilon_{kk}^c \delta_{ij} + 2\mu \varepsilon_{ij}^c)\varepsilon_{ij}^c \quad (i, j = 1, 2, 3, \ldots)$$

(16)

3D FEM simulation features, excavation phase

Fig. 2 shows the model geometry used for 3D FEM simulations. The size of model is 25 m, 50 m and 50 m along the direction perpendicular to the tunnel axis, the direction of an axis of the tunnel and depth, respectively. The section of the tunnel is 7.5 m high and 2.5 m wide, and the first and second boreholes are 1.75 m in diameter and 6.5 m in depth. As a boundary condition, displacement at the outer boundaries of the model region is fixed in the normal direction and the initial principal stresses (in situ stresses), $\sigma_1 = -30.0$ MPa, $\sigma_2 = -15.0$ MPa and $\sigma_3 = -10.0$ MPa, are prescribed. The directions of σ_1, σ_2 and σ_3 are defined as the x-, z- and y-direction, respectively. It should be noted that negative stress values represent compression. The measuring point A is located at the depth of 1.95 m and 0.003 m into the pillar wall (DQ0063G01 shown in Fig. 2) in the narrowest part of the pillar.

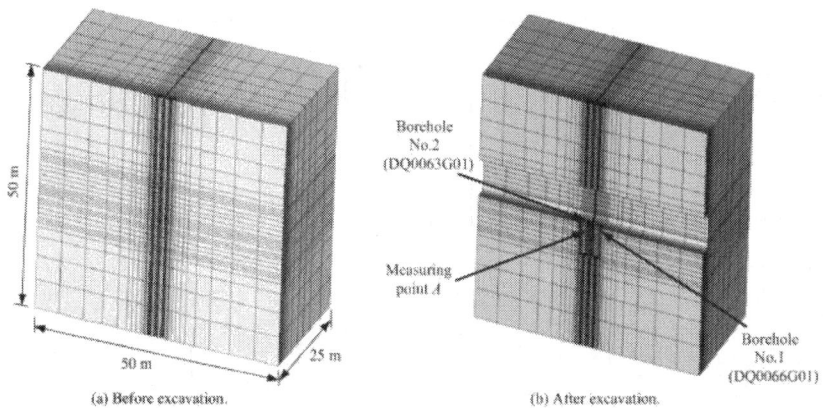

Figure 2. Model geometry for 3D FEM simulations.

The analysis steps of the excavation phase in the 3D simulations is listed in Table 1.

Table 1. Analysis steps of the excavation phase in the 3D simulations.

Step No.	Excavation phase	Step No.	Excavation phase
1	Excavation of access tunnel	11	Borehole No. 1 is confined at 0.7 MPa
2	Borehole No. 1, 0.5 m deep	12	Borehole No. 2, 0.5 m deep
3	Borehole No. 1, 1.0 m deep	13	Borehole No. 2, 1.0 m deep
4	Borehole No. 1, 2.0 m deep	14	Borehole No. 2 2.0 m deep
5	Borehole No. 1, 2.5 m deep	15	Borehole No. 2, 2.5 m deep
6	Borehole No. 1, 3.0 m deep	16	Borehole No. 2, 3.0 m deep
7	Borehole No. 1, 4.0 m deep	17	Borehole No. 2, 4.0 m deep
8	Borehole No. 1, 5.0 m deep	18	Borehole No. 2, 5.0 m deep
9	Borehole No. 1, 6.0 m deep	19	Borehole No. 2, 6.0 m deep
10	Borehole No. 1, 6.5 m deep	20	Borehole No. 2, 6.5 m deep

The number of nodes and elements are 27,744 and 24,800, respectively. The damage parameters used in the simulation were determined by fitting the stress-strain curve from uniaxial compression tests with radial strain control performed by SKB (Staub et al., 2004). The identified damage parameters are summarized in Table 2.

Table 2. Parameters for 3D FEM simulation.

Damage parameters					Hydraulic property, k_0 (m/s)
n_v	n_d	K_v (kPa)	K_d (kPa)	B_0 (kPa)	
0.7	0.7	61,715.00	942	93	5.0×10^{-8}

Mechanical property		Thermal property		
Young's modulus, E_0(GPa)	Poisson's ratio, v_0	Thermal conductivity, λ(W/(m K))	Specific heat, C(MJ/(m³ K))	Coefficient of thermal expansion, as (K^{-1})
72	0.25	2.6	2.1	7.0×10^{-6}

3D FEM simulation features, heating phase

After the second borehole is excavated, heating is carried out. The heated length is 6.5 m. During heating, the first borehole is filled (applying 0.7 MPa confining pressure). The displacement at the outer boundaries of the model region is fixed in the normal direction. The temperature is fixed at the top and bottom boundaries at the initial temperature (14.5 °C). A convective boundary condition, where the heat transfer coefficient is 4.65 W/(m² K) and reference temperature is 14.5 °C, is prescribed to the boundaries along the tunnel wall and the second borehole. Other boundaries (including the wall of the first borehole) are adiabatic. The locations of heaters and temperature monitoring points are shown in Fig. 3, where DQ0063G01 is the first borehole and KQ0064G06 is the observation borehole (temperature monitoring point). Fig. 4 shows the heat power during heating phase (Andersson, 2007, Andersson and Martin, 2009 and Andersson et al., 2009). The stress values from the last step of the excavation phase were used as the initial stress values for heating analysis. The parameters for coupled TM analysis are summarized in Table 2.

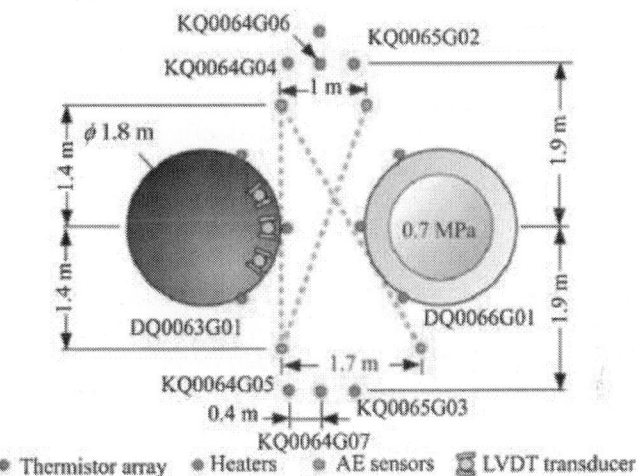

Figure 3. Layout of boreholes heaters and sensors (Andersson, 2007).

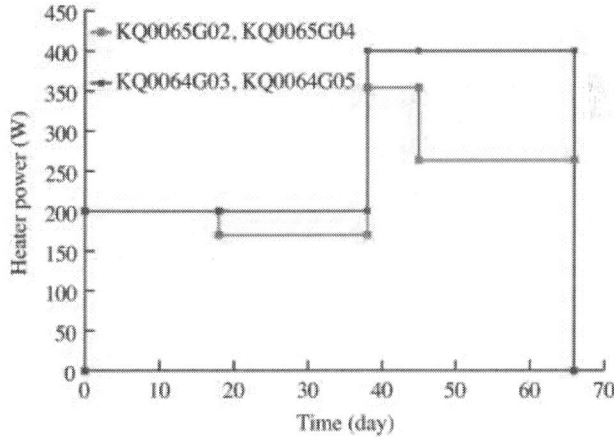

Figure 4. Heater power (Andersson, 2007).

Features of the 2D FEM simulations

A 2D FEM simulation was also performed to compare simulation results with 3D FEM and 2D DEM results. Fig. 5 shows the finite element mesh used for the 2D FEM simulation and mechanical boundary conditions. The displacement in the normal direction is fixed along the domain. The initial principal stresses σ_1 (in x-direction) and σ_3(in y-direction) are −44.37 MPa and −11.70 MPa, respectively.

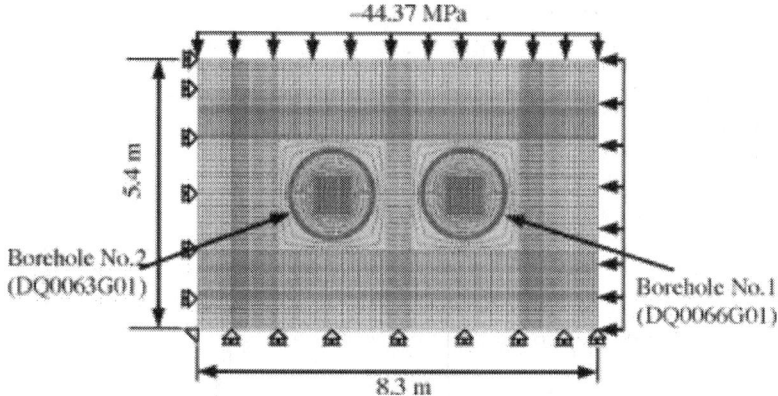

Figure 5. Simulation model for 2D FEM and mechanical boundary conditions.

The difference in initial in situ stresses is due to the excavation of access tunnel. These principal stresses are decided based on the results obtained from the 3D FEM simulation mentioned in the previous section. During the borehole excavation, the following four analysis steps were sequentially modeled to investigate the mechanical responses of the rock:

1) Step 0: excavation of the tunnel (initial condition).
2) Step 1: excavation of the borehole No. 1.
3) Step 2: 0.7 MPa confinement at borehole No. 1.
4) Step 3: excavation of the borehole No. 2.

As shown in Fig. 6, the outer boundaries of the domain and walls around the borehole No. 1 were assumed to be adiabatic, and the temperature around the borehole No. 2 was fixed at 14.5 °C. The thermal parameters used in 2D FEM simulations are the same as the ones used in the 3D simulations (see Table 2).

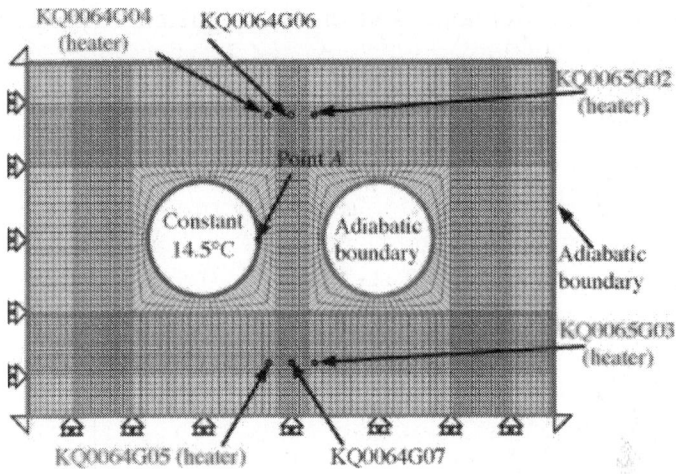

Figure 6. Thermal boundary conditions for heating process.

DISCONTINUUM BASED APPROACH, 2D DEM SIMULATIONS

Outline of the DEM with particles

The DEM with particles was originally investigated and pioneered by Cundall and Strack (1979). The calculations performed in the DEM can be described by the movement of each particle and the force/moment acting at each contact of the particles. A force–displacement law is used to determine the contact forces arising from the relative motion at each contact. The forces from all of the contacts on a particle are summed yielding a resultant force, and Newton's second law gives the translational as well as rotational motion of a particle resulting from the force acting on it. The new state of contacts is re-evaluated with the newly computed motion of particles, and a new cycle of computation is run. Although the DEM is originally a numerical technique for discontinuum, it becomes possible to be applied to a continuum by introducing bonding between particles (Potyondy and Cundall, 2004 and Shimizu et al., 2010). In this section, a summary of the formulation for the mechanical behavior of bonded particles and the newly introduced thermal flow algorithm to the original DEM code (Shimizu et al., 2010) are given.

Formulation for mechanics of bonded particles

In 2D DEM, the intact rock is modeled as a dense packing of small rigid circular particles. Neighboring particles are bonded together at their contact points with a set of three kinds of springs as shown in Fig. 7 and interact with each other. The normal force, f_n, the tangential force, f_s, and the moment, $f\theta$, are produced by the relative motion of the bonded particles (see Figs. 7a–c), and are given by

$$f_n = k_n(dn_j - dn_i) \tag{17}$$

$$f_s = k_s \left[ds_j - ds_i - \frac{L}{2}(d\theta_j + d\theta_i) \right] \tag{18}$$

$$f_\theta = k_\theta(d\theta_j - d\theta_i) \tag{19}$$

where k_n, k_s and $k\theta$ are the spring stiffness of normal, shear and rotational spring, respectively; dn and $d\theta$ are displacements in normal and shear directions and rotation of each particles, respectively.

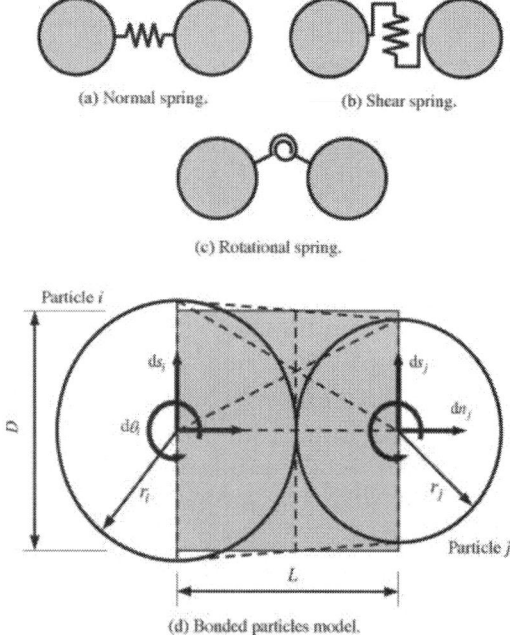

(a) Normal spring.   (b) Shear spring.

(c) Rotational spring.

(d) Bonded particles model.

Figure 7. Three different types of contact springs and bond between two bonded particles.

The stiffness of the normal and rotational springs, k_n and $k\theta$, are calculated using beam theory, and the stiffness of shear springs k_s is calculated by multiplying the stiffness of the normal spring k_n by a constant stiffness ratio a. Hence, the stiffness of the springs is given by the following equations:

$$k_n = \frac{E_p A}{L}$$

(21)

$$k_s = \alpha k_n$$

(22)

$$k_\theta = \frac{E_p I}{L}$$

(22)

where A is the cross-sectional area of the bond, I is the moment of inertia of the bond, and E_p is the Young's modulus of particle and bonds. The moment of inertia I depends on the shape of the cross-section, and rectangular cross-section is assumed in this study. A bond is shown schematically as a gray rectangle in Fig. 7d. In this figure, L and D are the bond length and bond diameter, respectively, and can be expressed as

$$L = r_i + r_j$$

(23)

$$D = \frac{4 r_i r_j}{r_i + r_j}$$

(24)

where r_i and r_j are the radii of the two bonded particles.

Since the DEM uses a fully dynamic formulation, small amounts of viscous damping are necessary to help in providing dissipation of high-frequency motion. If contact damping is not included, the assemblies would not be able to reach equilibrium. Contact damping operates on the relative velocities at the contacts and may be envisioned as resulting from dashpots acting in the normal and shear directions at the contacts. The coefficients of viscous contact damping in the normal and shear directions are represented by C_n and C_s, respectively:

$$C_n = 2\sqrt{m_{ij} k_n}$$

(25)

$$C_s = C_n \sqrt{\frac{k_s}{k_n}}$$

$$(26)$$

where mij can be given by the product of the weight of two particles mi and mj:

$$m_{ij} = 2\frac{m_i m_j}{m_i + m_j}$$

$$(27)$$

Crack generation mode and bond breakage of the particles
If the normal stress, σ, exceeds the strength of the normal spring, S_t, or shear stress, τ, exceeds the strength of the shear spring, S_s, then the contact bond breaks, and the three springs are removed from the model. A rotation restriction spring is used only for the force calculation and no breakage criterion is applied to this spring. Each bond breakage represents a microcrack.

1) Normal spring breakage criterion:

$$\left.\begin{array}{ll} \tau < 0 & \text{(Tensile stress)} \\ \tau \geq S_t & \text{(Bond break)} \end{array}\right\}$$

$$(28)$$

2) Shear spring breakage criterion:

$$|\tau| \geq S_s \text{(Bond break)}$$

$$(29)$$

In this study, the crack modes are classified by the shear–tensile stress ratio $|\tau/\sigma|$ regardless of broken spring type (normal or shear springs).

1) Crack mode criterion 1:

$$\left.\begin{array}{l} \sigma < 0 \quad \text{(Tensile stress)} \\ |\tau/\sigma| \leq 1 \end{array}\right\} \quad \text{(Tensile crack)}$$

$$(30)$$

2) Crack mode criterion 2:

$$\left.\begin{array}{l} \sigma < 0 \quad \text{(Tensile stress)} \\ |\tau/\sigma| > 1 \end{array}\right\} \text{(Shear crack)} \tag{31}$$

3) Crack mode criterion 3:

$$\sigma > \text{(Shear crack)} \quad 0 \text{ (Compressive stress)} \tag{32}$$

When a microcrack is generated and dissipated to neighboring springs, the strain energy stored in both the normal and the shear springs at the contact point is released. For the DEM simulations presented in this paper, the strain energy is calculated using the following equation, which represents the energy of acoustic emission (AE):

$$E_k = \frac{f_n^2}{2k_n} + \frac{f_s^2}{2k_s} \tag{33}$$

When the unbonded particles and/or particles with bond breakage are in contact, springs and dashpots are inserted into the contact point in both the normal and the tangential directions, and compressive normal force f_n and tangential (frictional) force f_s act at the contact points. The normal-direction springs satisfy a non-tension constraint.

2D DEM simulation features, excavation phase

The model geometry for 2D DEM simulation is shown in Fig. 8. The size of the model is 8.3 m in the direction of the tunnel axis (x-direction), and 5.4 m in the direction of the tunnel width (y-direction). Only horizontal 2D sections are modeled in this study (at the depth of 3.5 m from the tunnel floor). The rock model is expressed by the assembly of bonded particles. Particles are irregularly and randomly arranged and the number of particles is 217,367 with the particle radius ranging from 5 mm to 10 mm. The particle radii were determined to satisfy a uniform distribution between maximum and minimum particle radius. As shown in Fig. 7, two boreholes (Nos. 1 and 2) are excavated to form a pillar area with a width of 1.0 m at the thinnest part. The diameter of both boreholes is 1.8 m. The monitoring point A is located on the surface of borehole No. 2.

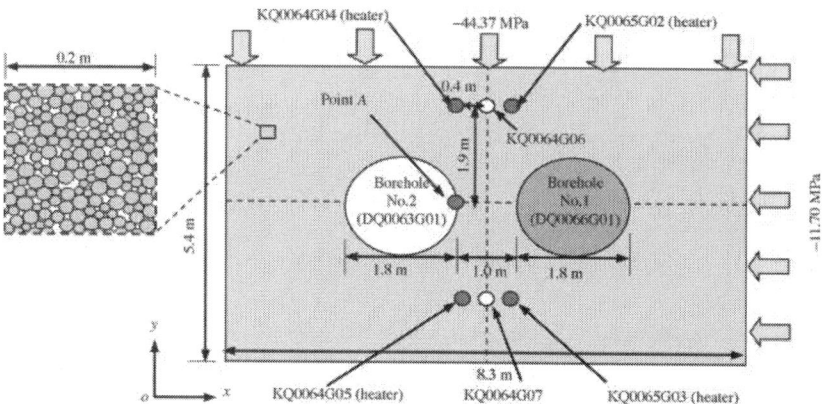

Figure 8. Model geometry and loading condition for pillar experiments using 2D DEM. Particles are irregularly and randomly arranged with the particle radius ranging from 5 to 10 mm.

The rock model is surrounded by the four confining walls. The normal displacement at left and bottom walls is fixed and a constant confining pressure (in situ stresses) is applied at the right and upper walls. The stress applied to the rock model is calculated from the width and height of the model and total force acting on each loading platen from particles. Two confining stresses of −44.37 MPa and −11.7 MPa are applied to the rock model in the x- and y-directions, respectively. These confining stresses are the same as the principal stresses applied in the 2D FEM simulations. Frictional force between the rock model and the confining walls is not considered. The analysis steps are the same as those used in 2D FEM simulation.

Input microscopic mechanical parameters for the 2D DEM simulations were calibrated to match the macroscopic mechanical properties of the Äspö diorite (Staub et al., 2004). The laboratory results from uniaxial/triaxial compression tests and Brazilian tensile test were used. The calibrated parameters are summarized in Table 3.

Table 3. Rock model properties and calibration results for 2D DEM simulation.

Rock model data										
Width of rock model (m)	Height of rock model (m)	Number of particles	Maximum particle radius (mm)	Minimum particle radius (mm)	Particle density (kg/m³)	Friction coefficient of wall, $\tan\phi_w$	Poisson's ratio of wall, ν_w	Young's modulus of wall, E_w (GPa)	Friction coefficient of particle, $\tan\phi_p$	Poisson's ratio of particle, ν_p
8.3	5.4	217,387	10	5	2741	0.0	0.3	200	0.5	0.27

Tuning parameters				Calibration results							
Young's modulus of particle, E_p (GPa)	Shear/normal spring stiffness ratio, α	Shear strength of bonding, τ_s (MPa)	Tensile strength of bonding, σ_t (MPa)	UCS of rock model (MPa)		Young's modulus of rock model (GPa)		Poisson's ratio of rock model		Tensile strength of rock model (MPa)	
				Experiment	Simulation	Experiment	Simulation	Experiment	Simulation	Experiment	Simulation

2D DEM simulation features, heating phase

For the TM analysis, the thermal flow algorithm was newly introduced to the original DEM code. When two particles with different temperatures are in contact, the rate of heat flow between these two particles can be given by the following equation based on the Fourier's law (see Fig. 9a):

$$Q = kA' \frac{\Delta T}{L} \tag{34}$$

where Q is the heat flow, k is the thermal conductivity of the particles, A' is the area of the contact point, ΔT is the temperature difference between the two particles, and L is the distance between the centers of the two particles.

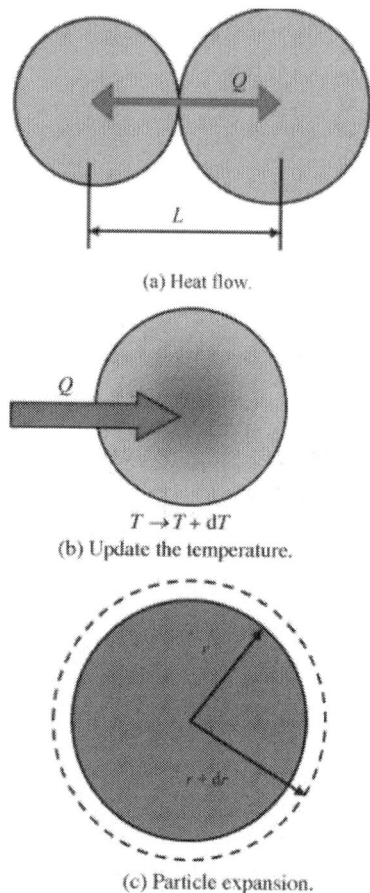

(a) Heat flow.

$T \rightarrow T + dT$
(b) Update the temperature.

(c) Particle expansion.

Figure 9. Thermal flow algorithm for DEM model.

The temperature of a particle is updated during the heat calculation (see Fig. 9b). The change in temperature of a particle, dT, is given by

$$dT = \frac{Q dt}{Cm}$$

(35)

where C is the specific heat of the particle, m is the mass of the particle, and dt is the duration of the time step.

When the temperature of a particle is updated, the radius of the particle is modified (seeFig. 9c). The radius increment of the particle, dr, is given by

$$dr = a_T r dT \qquad (36)$$

where aT is the coefficient of thermal expansion of a particle.

The thermal properties of rock model for 2D DEM simulations during the heating phase are listed in Table 4. These thermal properties were determined from the results obtained from the pillar stability experiments reported in the SKB report (Andersson, 2007). An adiabatic boundary condition was applied along the borehole No. 1 and the outer border of domain. The temperature was fixed at 15 °C along the borehole No. 2. The location of heaters and temperature monitoring points are shown in Fig. 8.

Table 4. Parameters for thermal flow (Andersson, 2007).

Thermal conductivity, k(W/(m K))	Thermal capacity, C(MJ/(m³ K))	Coefficient of thermal expansion, aT(K^{-1})
2.6	2.1	7.0×10^{-6}

SIMULATION RESULTS

Simulation results obtained using 3D FEM

The calibration of damage parameters was carried out using 2D FEM model. Table 5shows the calibration cases. Case 1-1 (see Table 2) is the basic case using the damage parameters estimated by fitting stress-strain curves from the laboratory test. In Cases 2 and 3, the effect of a change of the elastic modulus and damage parameter B_0 is examined, respectively. The effect of damage parameters K_d, K_v as well as n_d, n_v is also examined in Cases 4 and 5, respectively. In Case 6, the elastic modulus is decreased to 75% of the one obtained by the laboratory test and the value of B_0 is adjusted to zero. The other damage parameters are changed in the same rate (decreased to 30%, 20% and 10% of original values). In Case 7, the values of the elastic modulus and B_0 are the same as those in Case 6 and K_d, K_v are changed in the same decreasing rate but nd, nv are fixed at the original values. All the data adjusted are indicated in gray in Table 5.

Table 5. Calibration cases for damage parameters.

Case No.	E (GPa)	Damage parameters				
		B_0 (kPa)	K_d (kPa)	K_v (kPa)	n_d	n_v
1-1	72.0	93.0	942.0	61,715.0	0.70	0.70
2-1	36.0 (50%)	93.0	942.0	61,715.0	0.70	0.70
2-2	7.2 (10%)	93.0	942.0	61,715.0	0.70	0.70
3-1	72.0	0.0	942.0	61,715.0	0.70	0.70
4-1	72.0	93.0	471.0 (50%)	30,857.2 (50%)	0.70	0.70
4-2	72.0	93.0	94.2 (10%)	6171.5 (10%)	0.70	0.70
5-1	72.0	93.0	942.0	61,715.0	0.35 (50%)	0.35 (50%)
5-2	72.0	93.0	942.0	61,715.0	1.40 (200%)	1.40 (200%)
5-3	72.0	93.0	942.0	61,715.0	7.00 (1000%)	7.00 (1000%)
6-1	54.0 (75%)	0.0	282.6 (30%)	18,514.5 (30%)	0.21 (30%)	0.21 (30%)
6-2	54.0 (75%)	0.0	188.4 (20%)	12,343.0 (20%)	0.14 (20%)	0.14 (20%)
6-3	54.0 (75%)	0.0	94.2 (10%)	6,171.5 (10%)	0.07 (10%)	0.07 (10%)
7-1	54.0 (75%)	0.0	282.6 (30%)	18,514.5 (30%)	0.70	0.70
7-2	54.0 (75%)	0.0	235.5 (25%)	15,428.8 (25%)	0.70	0.70
7-3	54.0 (75%)	0.0	188.4 (20%)	12,343.0 (20%)	0.70	0.70

Table 6 shows the threshold of damage variable. When D is smaller than 0.1, the material is not damaged. Some fractures might be generated in the material when D is larger than 0.1, and the spalling might occur when D is larger than 0.3 (Chijimatsu et al., 2011). The calibration for elastic modulus and damage parameters was performed by comparing the distribution of damaged zone around the borehole No. 2 where the spalling was also observed during excavation and heating phases in APSE (Andersson, 2007). According to the simulations with different values for damage parameters, the parameter set for Case 7-3 showed better agreement with observed/measured spalling region (Chijimatsu et al., 2011).

Table 6. Judgment of damage value.

D	Judgment (empirical fact)
<0.1	No damage or no fracture
>0.1	Possibility of generation of fractures
>0.3	Possibility of generation of spalling

The TM behavior in APSE was re-examined using the parameters obtained from the above-mentioned calibration results (a set of damage parameters for Case 7-3 in Table 5). Fig. 10 shows the simulation results after the excavation of boreholes. The distribution of maximum principal stress and maximum tangential stress are shown in Fig. 10a and b, respectively. Fig. 10c and d shows the distribution of damage variable and hydraulic conductivity, respectively. The damage occurs in the pillar (the region between the boreholes) and the value of damage variable becomes 0.3 and/or more, which means that spalling might occur. The estimated depth of spalling region shows good agreement with the observation at site. The hydraulic conductivity in the pillar part is significantly increased by the excavation damage.

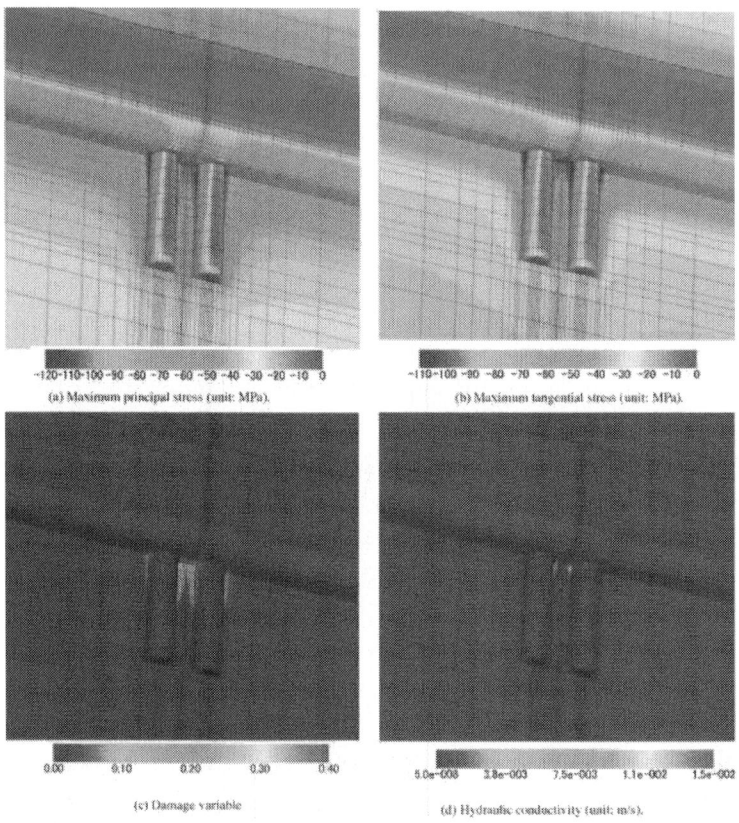

(a) Maximum principal stress (unit: MPa).

(b) Maximum tangential stress (unit: MPa).

(c) Damage variable

(d) Hydraulic conductivity (unit: m/s).

Figure 10. 3D FEM simulation results during excavation phase (Step 20).

Simulation results by DEM and their comparison with FEM
Excavation phase
Fig. 11 shows distribution of the maximum and minimum principal stresses (σ_1 and σ_3) as well as maximum tangential stress ($\sigma_1 - \sigma_3$) after the excavation phase calculated by 2D DEM. It should be noted that the minus stress values in the figures represent compression. From Fig. 11, the maximum principal stress at the borehole surface across the y-axis decreases to almost 0 MPa after the excavation of each borehole. On the other hand, maximum principal stress at the borehole surface across the x-axis increases. The minimum principal stress at the borehole surface decreases and becomes tensile stress. In particular, tensile stress at the borehole surface across the x-axis is the highest. The maximum tangential

stress at the borehole surface across the x-axis is the highest. Fig. 12 and Fig. 13 show the distribution of maximum and minimum principal stresses (σ_1 and σ_3) as well as the maximum tangential stress ($\sigma_1 - \sigma_3$) after the excavation phase calculated by 2D and 3D FEM (THAMES code), respectively. These results are similar to the ones obtained from 2D DEM simulations.

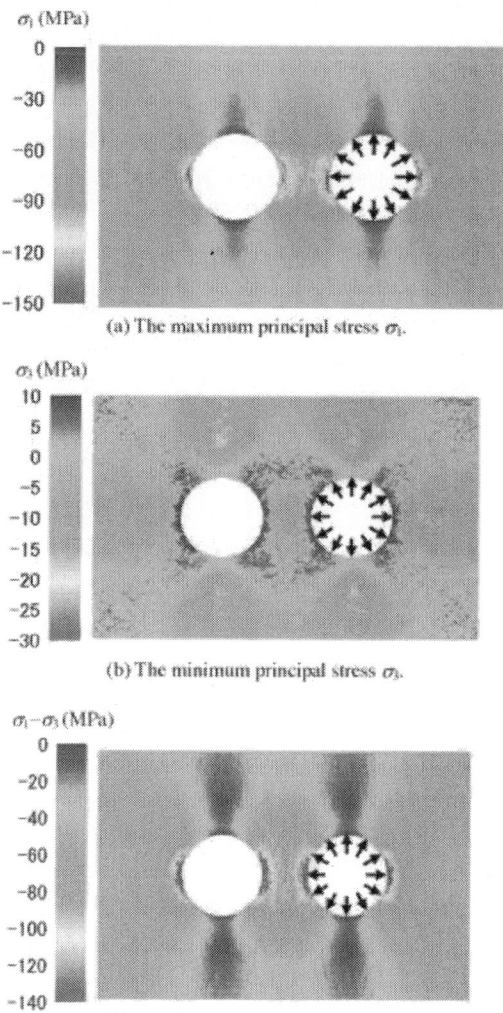

(a) The maximum principal stress σ_1.

(b) The minimum principal stress σ_3.

(c) The maximum tangential stress $\sigma_1 - \sigma_3$.

Figure 11. Distribution of the maximum and minimum principal stresses (σ_1 and σ_3) as well as the maximum tangential stress ($\sigma_1 - \sigma_3$) after excavation of the second borehole calculated by 2D DEM. Minus stress value represents compression.

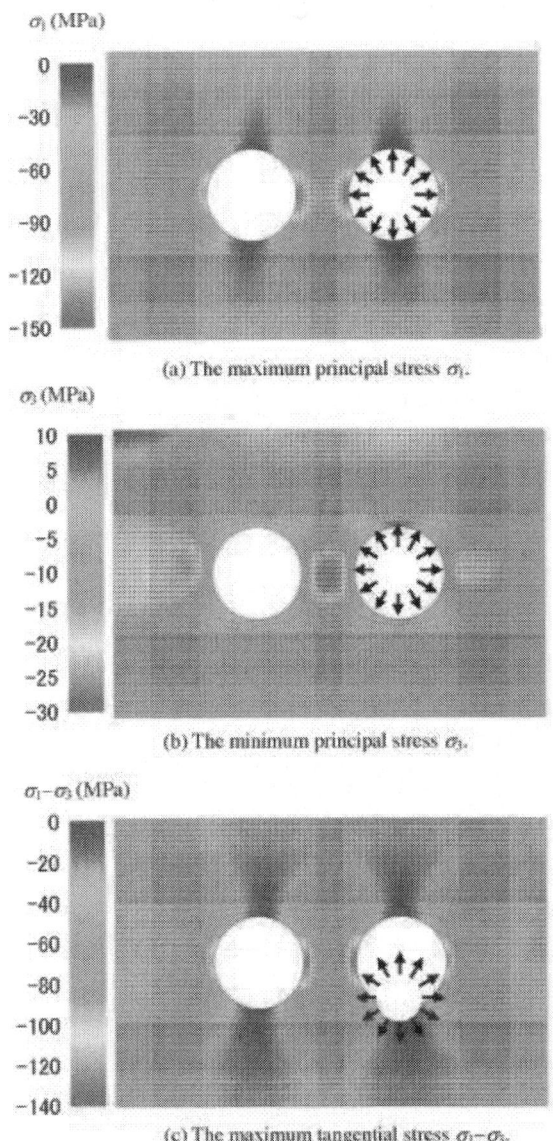

(a) The maximum principal stress σ_1.

(b) The minimum principal stress σ_3.

(c) The maximum tangential stress $\sigma_1 - \sigma_3$.

Figure 12. Distribution of the maximum and minimum principal stresses (σ_1 and σ_3) as well as maximum tangential stress ($\sigma_1 - \sigma_3$) after excavation of the second borehole calculated by 2D FEM. Minus stress value represents compression.

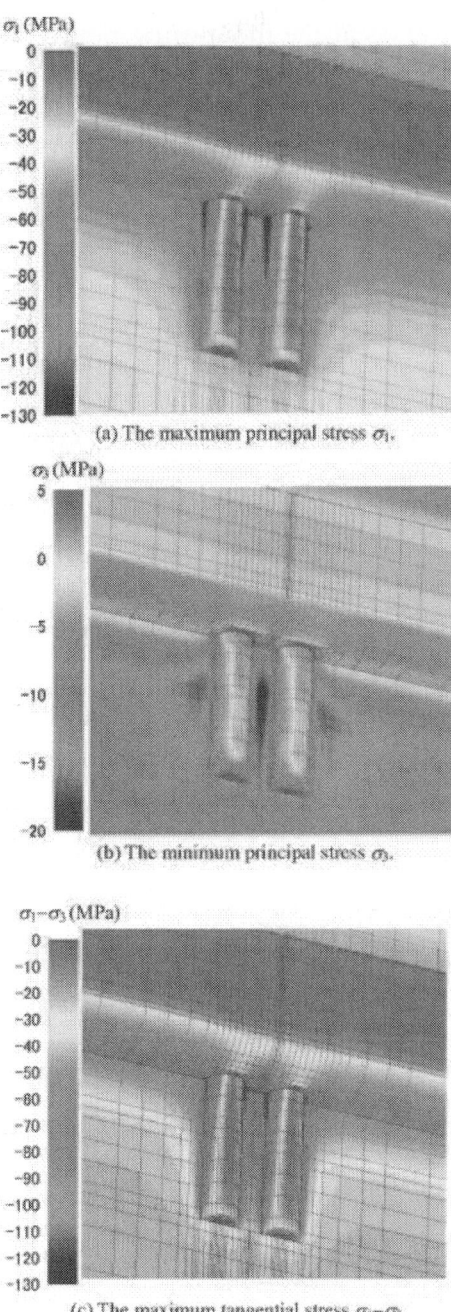

(a) The maximum principal stress σ_1.

(b) The minimum principal stress σ_3.

(c) The maximum tangential stress $\sigma_1 - \sigma_3$.

Figure 13. Distribution of the maximum and minimum principal stresses (σ_1 and σ_3) as well as maximum tangential stress ($\sigma_1 - \sigma_3$) after excavation of the second borehole calculated by 3D FEM.

The calculated stresses at the monitoring point A (see Fig. 8 for the location) for each excavation phase are summarized and compared in Table 7. The change of the maximum and minimum principal stresses (stress paths) at the monitoring point A for each excavation step calculated by 2D DEM are also compared with the ones obtained from 2D/3D FEM simulations and shown in Fig. 14. Fig. 15 also shows the measured variation of the maximum and minimum principal stresses at the monitoring point A during excavation. From these figures, the minimum principal stress calculated by 2D DEM is smaller than the one calculated by 2D/3D FEM and the one measured at site. This may be caused by the given mechanical boundary conditions applied in 2D DEM. In the DEM simulations, the left and bottom walls are fixed and constant confining pressures are applied at the right and top walls. The confining stresses from the wall are transmitted through the contact and/or bond of particles. However, the stress paths show similar trend, especially the stress path obtained from 3D FEM simulation agrees well with the one measured at site by SKB. The maximum tangential stress at monitoring point A increases drastically during/after excavation of borehole No. 2. The maximum tangential stress and its evolution calculated by 2D DEM and 2D/3D FEM show good agreement, even though the stress value calculated by 2D DEM is slightly large (see Fig. 16).

Table 7. The comparison of maximum and minimum principal stresses as well as maximum tangential stress at monitoring point A and their evolution during each excavation step between 2D DEM and 2D/3D FEM calculations. The stress values in brackets are obtained from 2D/3D FEM.

Analysis step	Maximum principal stress,σ_1 (MPa)	Minimum principal stress,σ_3 (MPa)	Maximum tangential stress,$\sigma_1 - \sigma_3$ (MPa)
Step 0	−49.9 (−44.4/−44.4)	−8.3 (−11.7/−11.7)	−41.6 (−32.7/−32.7)
Step 1	−60.6 (−55.6/−51.6)	−14.1 (−18.6/−14.2)	−46.5 (−37.0/−37.4)
Step 2	−60.4 (−55.5/−51.6)	−14.2 (−18.6/−14.2)	−46.1 (−36.8/−37.4)
Step 3	−148.4 (−149.9/−126.3)	−2.5 (−3.5/−6.0)	−145.9 (−146.4/−120.4)

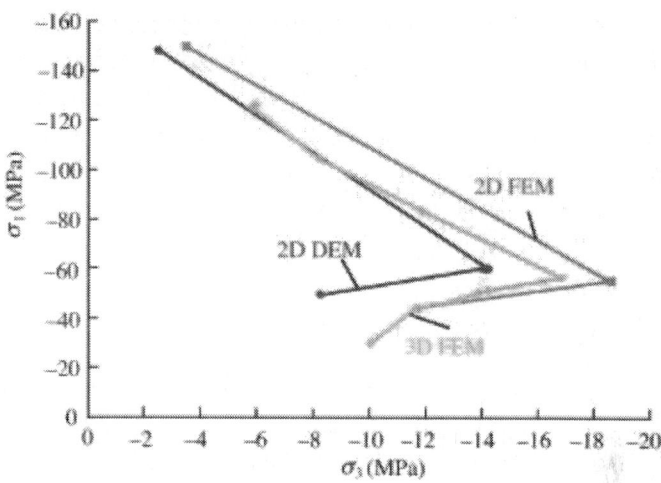

Figure 14. Comparison of the stress paths at point A between 2D DEM and 2D/3D FEM. Minus values mean compression.

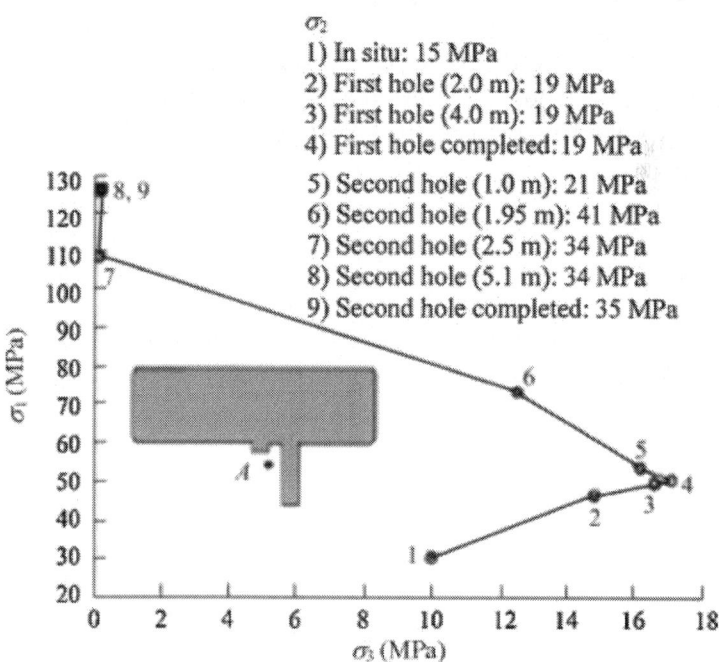

Figure 15. The stress paths at point A measured at site (Andersson, 2007). It should be noted that positive stress values represent compression.

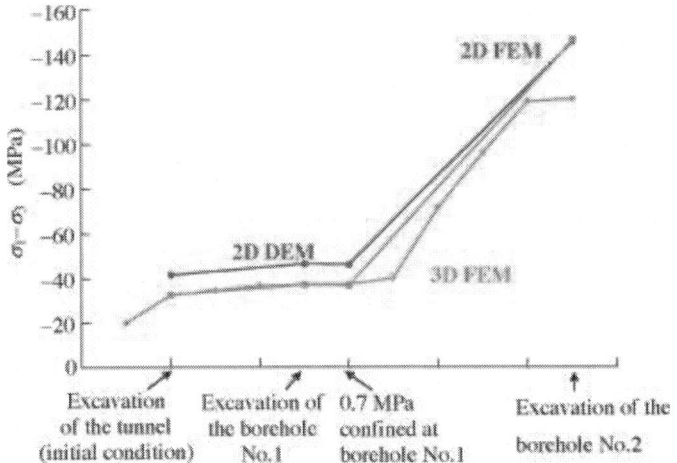

Figure 16. Comparison of the maximum tangential stress at point A between 2D DEM and 2D/3D FEM. Minus values mean compression.

Heating phase
Fig. 17 shows the distribution of temperature for 10, 20, 30, 40, 50, 70 days calculated by 2D DEM with thermal flow algorithm. As shown in Fig. 17, the temperature of the rock model increases with increasing heat output and the temperature decreases gradually after heating stop. A significant temperature increase can be observed only around the heater position. Fig. 18 and Fig. 19 also show the distribution of temperature for 10, 20, 30, 40, 50, 70 days calculated by 2D and 3D FEM (THAMES code), respectively.

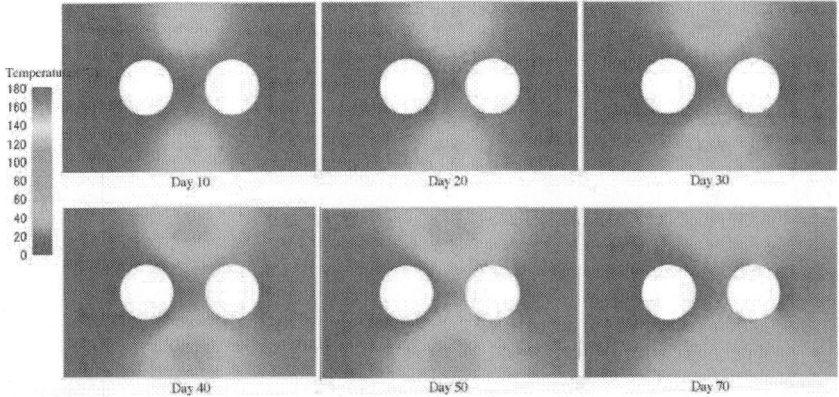

Figure 17. Temperature distribution for 10, 20, 30, 40, 50, 70 days (2D DEM).

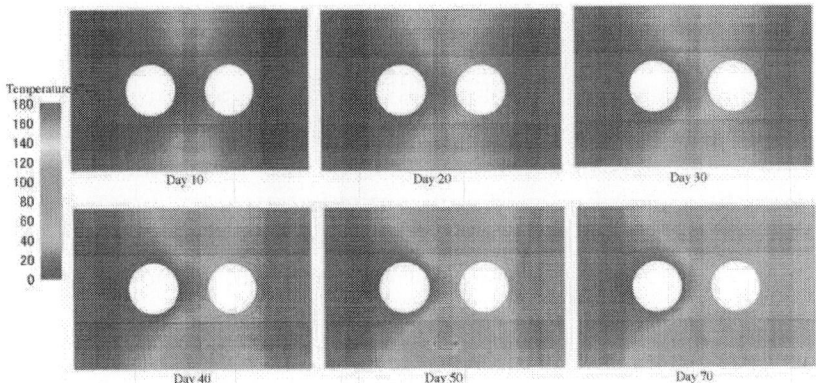

Figure 18. Temperature distribution for 10, 20, 30, 40, 50, 70 days (2D FEM).

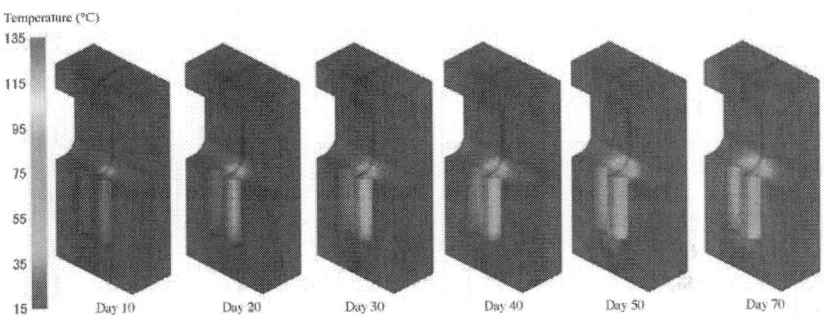

Figure 19. Temperature distribution for 10, 20, 30, 40, 50, 70 days calculated by 3D FEM (Chijimatsu et al., 2011).

Fig. 20 shows the comparison of temperature evolution at the monitoring point KQ0064G06 calculated by both 2D DEM and 2D/3D FEM. The evolution of measured temperature at site is also shown in Fig. 21. It should be noted that the 2D cross section at the depth of 3.5 m was selected to compare the results. From Fig. 20 and Fig. 21, overall trend of three calculated temperature evolution agrees well with the one monitored at site. Especially, the calculated temperature evolution by 3D FEM agrees quantitatively and qualitatively well with the one observed at site. However, the temperature calculated by 2D DEM and 2D FEM is higher than the experimental data reported by SKB (Andersson, 2007). This may be caused by the thermal boundary conditions in the model, which is adiabatic boundary, applied along the outer boundaries for the analytical domain and borehole No. 2. The temperature tends to be higher because the heater location is close to these boundaries. The plane strain approximation in 2D model also affects the calculated temperature distribution significantly because in 2D model the perpendicular heat flux in the depth direction is not considered. Moreover, the rock mass in 2D DEM is modeled as the assembly of many particles and the porosity of the model is larger than that of the real rock mass. Therefore, calibration of laboratory experiments (e.g. uniaxial/triaxial compressive tests) considering temperature will be necessary to investigate the effect of temperature on the microscopic parameters of particles for more realistic DEM simulations.

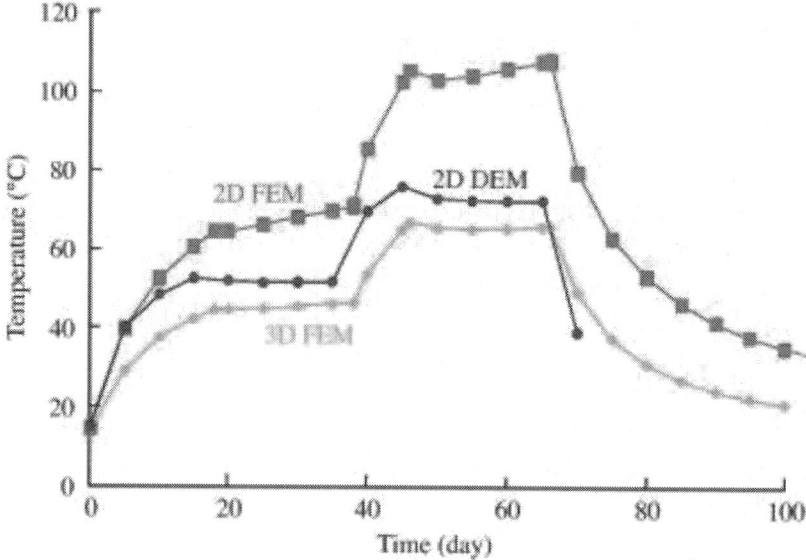

Figure 20. Comparison of calculated temperature evolutions at the monitoring point (KQ0064G06) between 2D DEM and 2D/3D FEM.

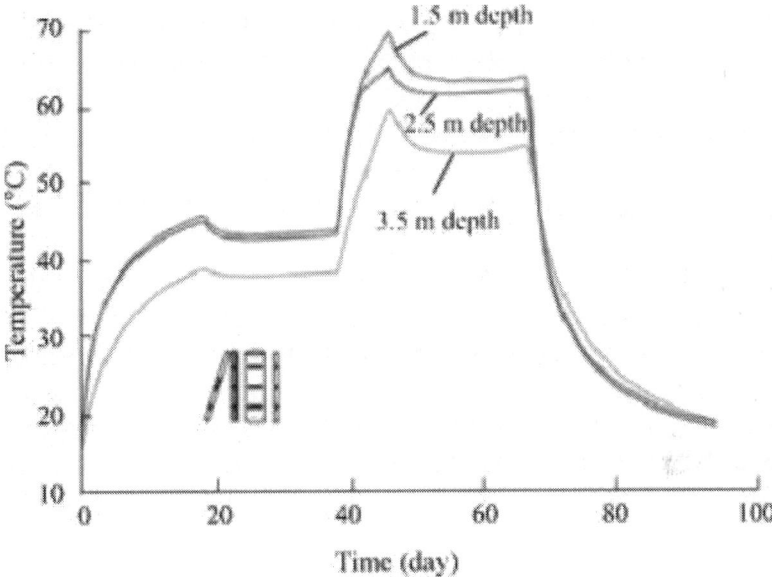

Figure 21. Temperature evolutions at monitoring point at site. Simulation results should be compared at the depth of 3.5 m (Andersson, 2007).

Fig. 22 shows the comparison of incremental thermal-induced maximum tangential stress at monitoring point A and its evolution during heating between 2D DEM and 2D/3D FEM. The maximum tangential stress at monitoring point A increases due to the expansion with increase of the temperature of the heated zones. The evolution of measured thermal-induced maximum tangential stress monitored at site is also shown inFig. 23. These simulation results agree qualitatively well with the experimental results. However, the thermal-induced maximum tangential stress calculated by 2D DEM is lower than the other results.

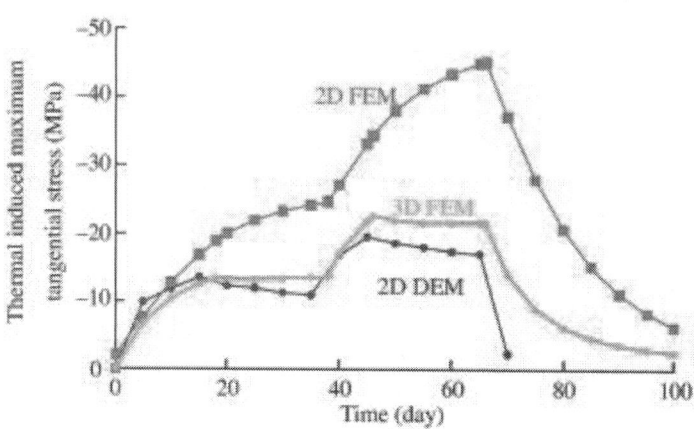

Figure 22. Comparison of incremental thermal-induced maximum tangential stress at monitoring point *A* and its evolution during heating between 2D DEM and 2D/3D FEM. Minus values represent compression.

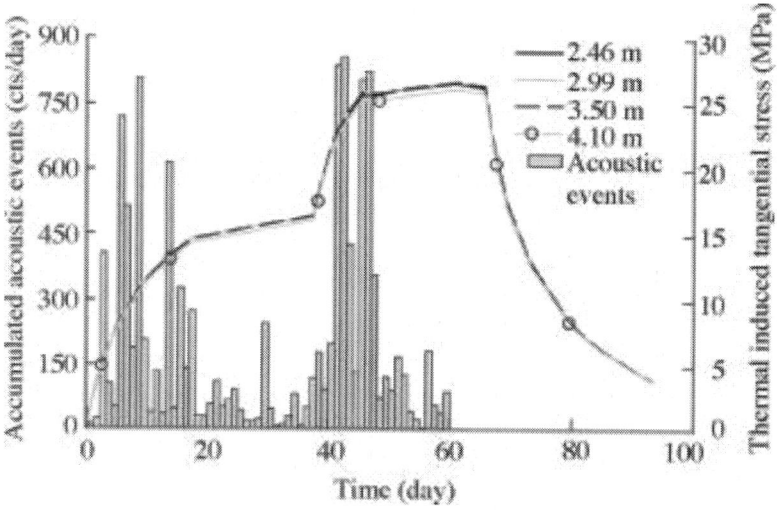

Figure 23. Evolution of thermal-induced maximum tangential stress at monitoring point *A* during heating and the associated daily acoustic events observed in the experiment (Andersson, 2007).

In the APSE, exfoliation of rock surfaces so-called spalling phenomena was observed around the borehole during heating (Andersson, 2007). Since DEM with particle is a discontinuum based numerical technique, generation and propagation of microcracks can be treated easily (not like FEM, which usually requires complicated re-mesh and/or adaptive mesh to simulate fracture generation and propagation). Fig. 24 shows the distribution of generated microcracks during excavation and heating phases. In Fig. 24, the microcracks generated during excavation and heating phases are indicated in black and red solid lines, respectively. New microcracks were generated around the borehole and gradually propagated with increasing temperature during heating phase. However, the number of microcracks was a few and spalling phenomena observed in the in situ experiment was not observed in the 2D DEM simulation, even if the calculated maximum temperature was higher than the experimental data. This may be caused by the relatively high microscopic parameters (especially strength for the contact bonds) calibrated from the laboratory uniaxial/triaxial compression tests using rock core sample. The particle size and its distribution may also significantly affect the fracture generation.

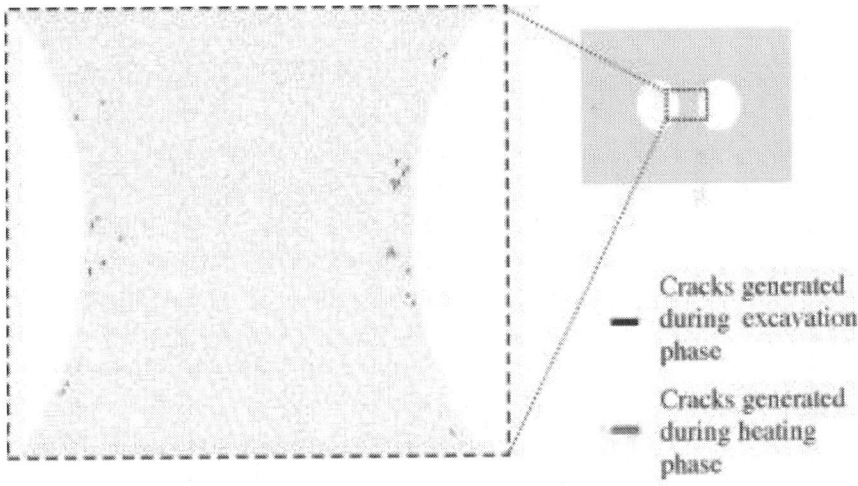

Cracks generated during excavation phase

Cracks generated during heating phase

Figure 24. Distribution of generated cracks during excavation and heating phases.

Fig. 25 shows incremental thermal-induced maximum tangential stress at monitoring point A and the number of generated microcracks per day during heating. From Fig. 25, the maximum tangential stress at monitoring point A increases with increasing heat output and temperature of rock mass. The number of generated microcracks increases with increasing maximum tangential stress at point A. This result agrees qualitatively well with the AE measurement at site as shown in Fig. 23 (Andersson, 2007).

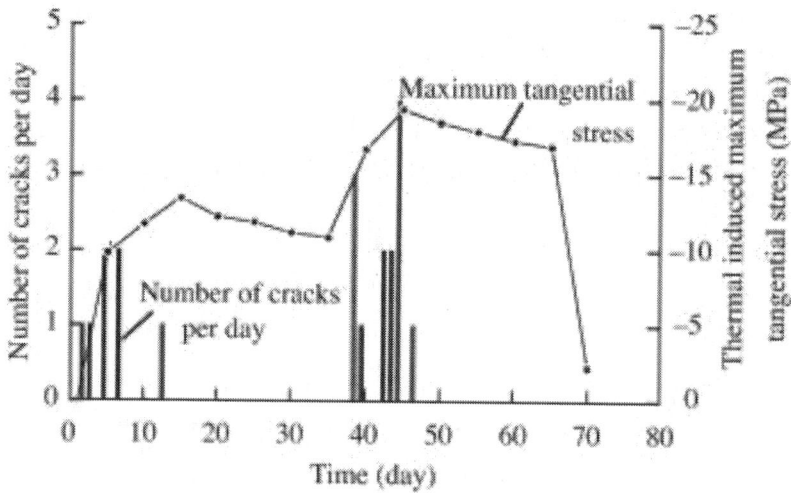

Figure 25. Evolution of maximum tangential stress around the monitoring point A and the number of microcracks per day.

Fig. 26 shows the displacement vectors of particles around the pillar during heating phase. The displacement of each particle oriented parallel to the radial direction from the heater position where the temperature highly increased and the particles near the monitoring point A moved into the borehole. The close-up view around the monitoring point A is shown in the right side in Fig. 26. Two microcracks, Cracks 1 and 2, are newly generated near the monitoring point A, and existence of these microcracks enlarges the displacement of the borehole surface.

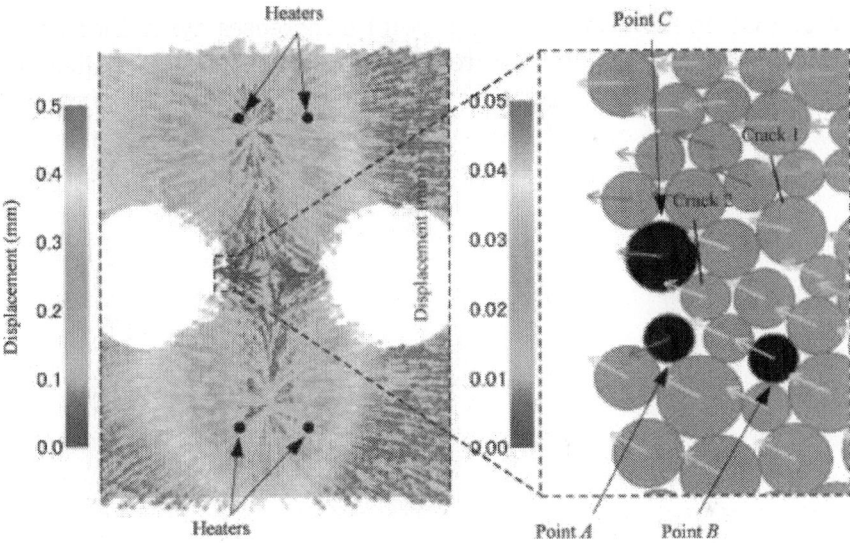

Figure 26. Displacement vectors of particles around boreholes and heaters during heating phase.

Fig. 27 shows the comparison of displacement at the measuring points A during heating calculated by 2D DEM and 2D/3D FEM. From this figure, the general trend of the displacement agrees for three different simulations. The borehole surface moves into the borehole when temperature of the rock mass increases. Especially, the displacement of the monitoring point A (see also Fig. 26) in DEM simulation is much larger. This is caused by newly generated microcracks around monitoring point A and particle can move easily. On the other hand, displacements of the points B and C in the 2D DEM simulation (close to monitoring point A (see also Fig. 26), but a little bit inside from the borehole surface) are comparable with 2D/3D FEM results. The displacement of both DEMs is smaller than the one calculated by 2D FEM. This may be caused by the fact that the calculated temperature by 2D DEM is lower than the one by 2D FEM.

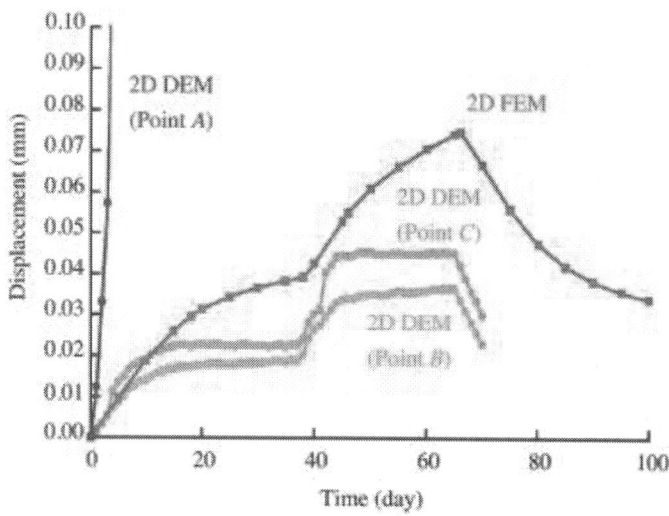

Figure 27. Comparison of displacement at monitoring point *A* between 2D DEM and 2D FEM simulations.

CONCLUSIONS

The coupled TM processes in the APSE were simulated using 2D and 3D FEM and 2D DEM with particles. The particle mechanics approach can treat the rock heterogeneity easily and investigate the failure mechanism in detail, including crack initiation and propagation. The thermal flow algorithm was newly introduced into the original DEM code to investigate the failure mechanism during not only excavation but also heating processes. The calculation results such as stress distribution, displacements as well as temperature distribution were compared with those monitored/measured at site and simulated by 2D and 3D finite element code called THAMES. The findings obtained from this study can be summarized as follows:

1) The calculated stress distribution around borehole and stress path at a point near the borehole wall by 2D DEM with particles agree qualitatively well with the ones obtained from 2D/3D FEM as well as in situ experiments at site.

2) The simulation results by 3D FEM show quantitatively good agreement with the data obtained from the measurements.

3) The newly developed thermal algorithm in 2D DEM produces similar temperature distribution to the one from 2D FEM.
4) The simulated crack generation and propagation during the excavation, pressurizing and heating processes by 2D DEM with particles agree qualitatively well with the observations at site. However, the number of cracks was few and spalling phenomena observed at site could not be captured sufficiently.
5) By comparing the simulation results obtained from two different approaches, quantitative and qualitative insights into various aspects of the processes occurring in the near field could be obtained.

ACKNOWLEDGEMENTS

The work described in this paper was conducted within the context of the international DECOVALEX Project (DEvelopment of COupled models and their VALidation against EXperiments). The work was financed by Japan Atomic Energy Agency (JAEA) who was also one of the Funding Organizations of the project. The views expressed in the paper are, however, those of the authors and are not necessarily those of the JAEA and other Funding Organizations. The research co-operation was coordinated by Dr. Christer Andersson from Swedish Nuclear Fuel and Waste Management Co. (SKB), Sweden. The data used in this work were provided by SKB through its Äspö Pillar Stability Experiment project.

REFERENCES

1. Andersson JC, Martin CD, Stille H. The Äspö Pillar Stability Experiment. Part II. Rock mass response to coupled excavation-induced and thermal-induced stresses. International Journal of Rock Mechanics and Mining Sciences 2009;46(5):879–95.
2. Andersson JC, Martin CD. The Äspö Pillar Stability Experiment. Part I. Experiment design. International Journal of Rock Mechanics and Mining Sciences 2009;46(5):865–78.

3. Andersson JC. Äspö Pillar Stability Experiment final report: rock response to coupled mechanical thermal loading. Stockholm: Swedish Nuclear Fuel and Waste Management Co. (SKB); 2007.

4. Chijimatsu M, Fujita T, Kobayashi A, Nakano M. Experiment and validation of numerical simulation of coupled thermal, hydraulic and mechanical behaviour in the engineered buffer materials. International Journal for Numerical and Analytical Methods in Geomechanics 2000a;24(4):403–24.

5. ChijimatsuM, Fujita T, Kobayashi A, NakanoM. Calibration and validation of thermal, hydraulic and mechanical model for buffer material. JNC technical report JNC TW8400 98-017. Tokai: JNC; 1998.

6. Chijimatsu M, Fujita T, Sugita Y, Amemiya K, Kobayashi A. In-situ coupled thermo-hydro-mechanical experiment at Kamaishi mine. In: GeoEng2000: an international conference on geotechnical and geological engineering. Lancaster: Technomic Publishing Co. Inc.; 2000b.

7. Chijimatsu M, Koyama T, Kobayashi A, Shimizu H, Nakama S. Simulation of the spalling phenomena at the Äspö Pillar Stability Experiment by the coupled T-H-M analysis using the damage model. In: Proceedings of the 4th international conference on coupled T-H-M-C processes in geosystems: fundamentals, modelling, experiments and applications (GeoProc2011); 2011. p. 1–13.

8. Cundall PA, Strack ODL. A discrete numerical model for granular assemblies. Geotechnique 1979;29(1):47–65.

9. Kobayashi A, Ohnishi Y. Effects of non-linearity of material properties on the coupled mechanical-hydraulic-thermal behavior in rock mass. In: Collected Papers of Japan Society of Civil Engineers. Tokyo: Japan Society of Civil Engineers; 1986. p. 101–10.

10. Lemaitre J. A course on damage mechanics. Berlin: Springer; 1992. p. 210. Murakami S, Kamiya K. Constitutive and damage evolution equations of elasticbrittle materials based on irreversible thermodynamics. International Journal of Mechanical Sciences 1997;39(4):473–86.

11. Ohnishi Y, Shibata H, Kobayashi A. Development of finite element code for the analysis of coupled thermo-hydro-mechanical behaviors of a saturated-unsaturated medium. In: Tsang CF, editor. Coupled Processes Associated with Nuclear Waste Repositories. Orlando: Academic Press; 1987. p. 551–7.

12. Potyondy DO, Cundall PA. A bonded-particle model for rock. International Journal of Rock Mechanics and Mining Sciences 2004;41(8):1329–64.
13. Shimizu H, Koyama T, ChijimatsuM, Fujita T, Nakama S. Coupled thermo-mechanical simulations for HLW disposal tunnel using distinct element method. Journal of the Society of Material Science, Japan 2011;60(5):470–6.
14. Shimizu H, Koyama T, Ishida T, Chijimatsu M, Fujita T, Nakama S. Distinct element analysis for Class II behavior of rocks under uniaxial compression. International Journal of Rock Mechanics and Mining Sciences 2010;47(2): 323–33.
15. Staub I, Andersson JC, Magnor B. Äspö Pillar Stability Experiment, geology and mechanical properties of the rock mass in TASQ. Stockholm: Swedish Nuclear Fuel and Waste Management Co. (SKB); 2004. p. 81–4.

CITATION

T. Koyama, M. Chijimatsu, H. Shimizu, S. Nakama, T. Fujita, A. Kobayashi, Y. Ohnishi, Numerical modeling for the coupled thermo-mechanical processes and spalling phenomena in Äspö Pillar Stability Experiment (APSE), Journal of Rock Mechanics and Geotechnical Engineering, Volume 5, Issue 1, February 2013, Pages 58-72, ISSN 1674-7755, http://dx.doi.org/10.1016/j.jrmge.2013.01.001.

CHAPTER 7

Discrete Modeling of Rock Joints with a Smooth-Joint Contact Model

C. Lambert[1], C. Coll[2,]*

[1] Department of Civil and Natural Resources Engineering, University of Canterbury, Private Bag 4800, Christchurch 8140, New Zealand

[2] Golder Associates, Christchurch, New Zealand

ABSTRACT

Structural defects such as joints or faults are inherent to almost any rock mass. In many situations those defects have a major impact on slope stability as they can control the possible failure mechanisms. Having a good estimate of their strength then becomes crucial. The roughness of a structure is a major contributor to its strength through two different aspects, i.e. the morphology of the surface (or the shape) and the strength of the asperities (related to the strength of the rock). In the current state of practice, roughness is assessed through idealized descriptions (Patton strength criterion) or through empirical parameters (Barton JRC). In both cases, the multi-dimensionality of the roughness is ignored. In this study, we propose to take advantage of the latest developments in numerical techniques. With 3D photogrammetry and/or laser mapping, practitioners have access to the real morphology of an exposed structure. The derived triangulated surface was introduced into the DEM (discrete element method) code PFC3D to create a synthetic rock joint. The interaction between particles on either side of the discontinuity was described by a smooth-joint model (SJM), hence suppressing the artificial roughness introduced by the particle discretization. Shear tests were then performed on the synthetic rock joint. A good correspondence between strengths predicted by the model and strengths derived from well-established techniques was obtained for the first time. Amongst the benefits of the methodology is the possibility offered by the model to be used in a quantitative way for shear strength estimates, to reproduce the progressive degradation of the asperities upon shearing and to analyze structures of different scales without introducing any empirical relation.

INTRODUCTION

The presence of discontinuities is inherent to almost any rock mass and is a major contributor to strength and deformation of rock structures (natural or engineering). The characteristics of those discontinuities not only control structurally controlled failures but also greatly influence the shear strength of the rock mass. Being able to describe the structure of a rock mass is critical to an understanding of its potential behavior. The development of various mapping techniques leads to a higher level of confidence on crucial characteristics such as location, orientation and persistence from which stochastic discrete fracture network (DFN) representations of the rock fabric are developed (Dershowitz, 1995 and Rogers et al., 2007). Based on numerical methods, equivalent rock mass can be created and tested in order to characterize its constitutive behavior (Pierce et al., 2007, Pine et al., 2007 and Deisman et al., 2010). These approaches are now able to model the engineering responses of rock and rock masses using some basic measured properties of the rock and the rock mass geometry as inputs. Offering a wider spectrum of predictions than the classical empirically-based classification schemes (anisotropy, heterogeneous, etc.), the synthetic rock mass approach and equivalents (Pierce et al., 2007, Pine et al., 2007 and Deisman et al., 2010) are turning to be a step forward for rock mechanics practitioners. However, the question of the shear strength of the discontinuities is in many cases poorly addressed in engineering practice despite having a significant impact on the rock mass strength (Lambert, 2008).

The shear behavior of discontinuities is a combination of various complex phenomena and interactions, such as dilation, asperity failure, deformation and interaction. Direct shear tests on natural rock discontinuities quickly enhanced the influence of roughness on the mechanical behavior of discontinuities. Barton (1973) proposed to assess roughness with an empirical parameter, joint roughness parameter (JRC), from which the shear strength of the discontinuity can be established. Initially estimated by visual comparison with standard roughness profiles, correlations between JRC and various statistical parameters or fractal dimension were established (Tse and Cruden, 1979 and Carr and Warriner, 1989). More recently, laser scanner and photogrammetry were used to define the surface topography and estimate its roughness (Grasselli, 2001, Hans and Boulon, 2003 and Haneberg et al., 2007). The dependence of shearing on the location and distribution of the three-dimensional (3D) contact area was demonstrated (Gentier et al., 2000) and new constitutive relations were developed based on a general description of roughness (Grasselli and Egger, 2003). Laser scanning and 3D photogrammetry techniques were applied in the field (Fardin et al., 2004) for large-scale surface measurements. Asperity shape and distribution on a discontinuity can now be measured with a great detail and potentially incorporated in any analysis. However with the complexity of the interaction between the two walls, a complete analytical formulation remains a hard task. Since the first idealized "saw-tooth" description proposed by Patton (1966), various constitutive models were developed that accommodate effect of asperities (Barton and Choubey, 1977 and Saeb and Amadei, 1992) and their progressive degradation during shearing (Plesha, 1987, Hutson and Dowding, 1990, Lee et al., 2001 and Misra, 2002) to name a few. Despite being each time more advanced, these models still rely on empirical relations or simplified descriptions of the surface asperities.

In an attempt to address this problem, many authors used numerical tools to assess the shear strength of discontinuities. Two-dimensional DEM (discrete element method) simulations were first presented as they offer a provision for asperity degradation

(Cundall, 2000 and Lambert et al., 2004). They have been successfully used to investigate gouge formation and evolution upon shearing (Zhao et al., 2012 and Zhao, 2013). However these simulations were at this stage limited to qualitative observations. Hybrid FEM/DEM (Karami and Stead, 2008) and FEM (Giacomini et al., 2008) methods proved their ability to reproduce typical behavior of rock joints including dilation and asperity degradation. Using 3D DEM, Kulatilake et al. (2001) showed that realistic macroscopic friction (i.e. at the joint level) could be obtained combining very small particles at the joint interface and extremely low contact friction. However this approach appears to be not very practical for engineering purposes. No formulation is available to calibrate the micro-properties of the joint model material against a given macroscopic behavior and the macroscopic friction targeted was quite high (friction coefficient of 0.7). In the field, discontinuities often exhibit a much lower strength. The particle size required may hence increase the computational cost to unpractical levels. Park and Song (2009) performed numerical shear tests on standard roughness profiles using the DEM code, PFC3D. This work once again highlighted the current limitations of particulate description as the discrete nature of the medium can introduce an artificial roughness to the discontinuity. The apparent roughness of the numerical specimen is higher than the introduced roughness (i.e. the initial roughness of the introduced surface or profile). The consequence is a slight overestimation of the strength and most importantly unrealistic predictions of dilation. The later point can be of major importance as joint aperture controls fluid flow in the discontinuities (Hans and Boulon, 2003 and Buzzi et al., 2008). The recent development of a new contact model named "smooth-joint model" (SJM) (Pierce et al., 2007) in PFC3D where particles are allowed to slide past one another without over-riding one another was a major breakthrough to represent discontinuities as planar surfaces associated to a realistic behavior for structural defects. In this study, we propose to develop in PFC3D a synthetic rock joint where a digital representation of a surface is introduced and described as a series of SJMs. The mechanical behavior of the synthetic rock joint is then analyzed performing numerical direct shear tests.

DEM SIMULATIONS OF CONSTANT NORMAL STRESS SHEAR TESTS

The discrete element method

The commercially available PFC3D (Itasca, 2008) software package was used for the 3D DEM simulations presented here. Unlike continuum codes, materials are described in PFC3D as a discontinuous medium as a collection of spherical rigid particles. The particles displace independently of one another following Newton's second law and interact with each other through contact forces that are generated at each contact point. Rock and more generally cohesive materials are represented as a bonded particle assembly, adding parallel bonds to create a synthetic material. A parallel bond acts like a conceptual cementitious material between particles. It has a finite dimension defined as a fraction of the particle diameter, a tensile and shear strength and a normal and tangential stiffness. When the contact force exceeds either tensile or shear strength, the parallel bond breaks and a micro-crack forms between the particles. Micro-cracks can eventually coalesce as external loading is applied and form fractures that can split the material into clusters. The location and the failure mode of the cracks are recorded. A detailed description of contact and bond models is provided in the user manual (Itasca, 2008).

The mechanical response of such assemblies, observed at a macroscopic level, is an emergent property of the complex interactions between the particles. Input parameters of the bonded particle model are micro-properties, contact properties and bond strength, and are not measurable with conventional laboratory apparatus. They are calibrated through an iterative process. Once a particle size distribution has been selected, cylindrical particle assemblies are generated and unconfined compression tests are simulated varying micro-properties until the mechanical response of the synthetic material conforms to the mechanical properties (i.e. uniaxial compressive strength, UCS; Young's modulus; Poisson's ratio) of the physical material (measured in the lab). A detailed description of the calibration procedure can be found in Potyondy and Cundall (2004). Once properly calibrated, such bonded assemblies proved their ability to reproduce typical behavior of rock-like materials (Kulatilake et al., 2001 and Potyondy and Cundall, 2004).

Properties of the granite considered for the scope of this study are given in Table 1. The micro-properties were calibrated accordingly. Normal and shear stiffnesses for contact and parallel bonds have impact on elastic properties of the particle assembly whereas bond shear and normal strengths mainly control UCS values. Various studies by Cundall (2000), Kulatilake et al. (2001) and Park and Song (2009) illustrated the necessity to introduce low particle friction to reproduce the shearing behavior of fracture planes in cohesive materials. In this study, bond strengths were calibrated considering zero friction between particles ($\phi_p = 0°$).

Table 1. Target (laboratory) and calibrated (calculation) bulk properties of the granite.

Method	Uniaxial compressive strength (MPa)	Young's modulus (GPa)
Laboratory	142.5	48.4
Calculation	143.8	48.6

Besides Potyondy and Cundall (2004) showed that particle friction impacts mainly on the post peak behavior of bulk material with little effect on peak strength. The influence of ϕ_p will be discussed with more detail in Section 3.3. The result of the calibration is given in Table 2 and the emergent bulk properties of the synthetic material are listed in Table 1.

Table 2. Material micro-properties for the granite sample.

Particle properties				
Particle mean radius (mm)	Particle radius ratio	Particle contact modulus (GPa)	Particle normal to shear stiffness ratio	Particle friction coefficient
1.5	1.66	56.1	2.5	0.0

Parallel-bond properties						
Parallel-bond radius multiplier	Parallel-bond modulus (GPa)	Normal to shear stiffness ratio	Bond normal strength (MPa)		Bond shear strength (MPa)	
			Mean value	Standard deviation	Mean value	Standard deviation
1.0	56.1	2.5	191	19.1	191	19.1

Description of the interface

The interface morphology used in the simulations is based on a natural discontinuity in granite studied by Grasselli (2001). The surface is 140 mm × 140 mm and the maximum amplitude of the asperities is around 9 mm. Fig. 1 shows a general view of the surface. The 3D surface was triangulated using a Kriging gridding method with a horizontal spacing of 1.4 mm between the grid points (in X- and Y-directions). Ninety nine profiles along the sliding direction (X-direction) were extracted for which the coefficient Z_2 (root mean square of the first derivative of the profile) was estimated:

$$Z_2 = \sqrt{\frac{1}{(N-1)\Delta x^2} \sum_{i=1}^{n}(z_{i+1} - z_i)^2}$$

(1)

where z_i and z_{i+1} are the elevation of two consecutive grid points on a profile, N is the total number of grid points on a profile, and Δx is the horizontal spacing.

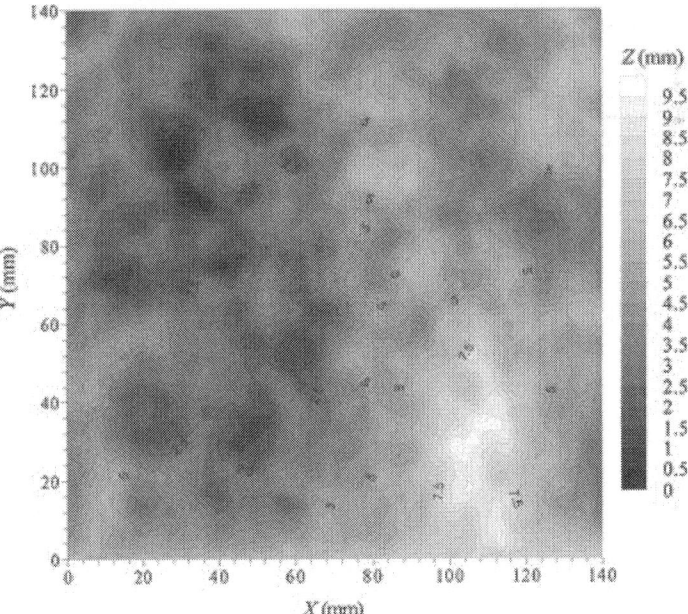

Figure 1. Morphology of the granite surface. All dimensions are in mm.

For each profile, a value of JRC was derived using the empirical relation proposed by Yang et al. (2001):

$$JRC = 32.69 + 32.98 \log_{10}(Z_2)$$ (2)

The profiles of the triangulated surface exhibited an average JRC of 10.4, varying from 4.9 to 13.9. Grasselli et al. (2002) estimated a JRC of 12.5 using a quantitative 3D surface description. The value obtained by Grasselli et al. (2002) is in the range of variation of JRC for individual profiles but appears to be slightly higher than the average value. However, the value of Z_2 and hence the derived JRC are sensitive to sampling intervals (Yu and Vayssade, 1991). Applying the same approach varying the horizontal spacing between the grid points of the triangulated surface, the average JRC increases to a value of 11.6 for a horizontal spacing of 0.56 mm (Lambert and Coll, 2009). The average JRC value of 10.4 was considered to be a reasonable estimate of the surface roughness used for this study.

The synthetic rock joint model
The numerical rock joint consists of a 140 mm × 140 mm × 50 mm (respectively X-, Y-, and Z-directions) parallelepiped particle assembly. The specimen genesis procedure is described in detail in Potyondy and Cundall (2004).

For the first series of simulations, three particle assemblies were generated, each one containing around 98,000 particles having a radius ranging from 0.5 mm (in the vicinity of the interface) to 2.4 mm. These specimens differ from one another only in their packing. A discontinuity is usually represented in PFC3D by debonding contacts along a surface. However, the particle geometry is still present and the discrete nature of the medium generates an artificial roughness that is added to one of the introduced surfaces, thus creating a particle size dependent joint behavior. For example, the DEM model presented by Park and Song (2009) using a standard profile with a JRC of 11.49 exhibited an apparent JRC of 17.55. To overcome the problem, an alternate scheme, termed as the "smooth-joint model" or SJM, initially proposed by Pierce et al. (2007), was implemented in PFC3D (Itasca, 2008). A smooth-joint model is a contact model that simulates the behavior of an interface, regardless of the local particle contact orientation along the interface. A typical smooth-joint is shown in Fig. 2. It allows particles to slide past one another without over-riding one another. A smooth joint is created by assigning this new contact model to all the contacts between particles that lie upon opposite sides of the surface. The SJM defines the tangential and normal directions according to the local orientation of the surface (by opposition to the initial normal and tangential directions of the contact, see Fig. 2). The joint normal and tangential force increments (ΔF_{nj} and ΔF_{tj}, respectively) are derived from normal and tangential displacement increments (ΔU_{nj} and ΔU_{tj}), multiplying by the joint stiffnesses ($\Delta F_{nj} = k_{nj}\Delta U_{nj}$ and $\Delta F_{tj} = k_{tj}\Delta U_{tj}$). The joint force is then adjusted to satisfy the force-displacement relationship and mapped back into the global system. This new formulation accommodates the standard shear behavior of a joint (friction, cohesion and dilation) independently of particle induced roughness. A complete description of the formulation can be found in the manual (Itasca, 2008). An initial study by Lambert et al. (2010) on the behavior of a rock-concrete interface suggested that realistic shear behavior, shear strength and dilation, could be obtained associating the SJM with a true morphology.

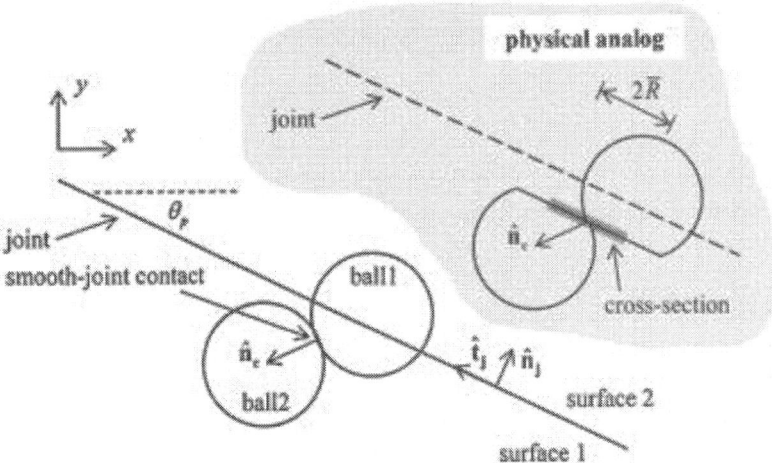

Figure 2. Smooth joint contact model between ball 1 and ball 2. Surface 1 and surface 2 denote either side of the joint lying at a dip angle of θ_p (after Itasca, 2008).

An algorithm was developed for the importation of the triangulated surface presented in Section 2.2 into a bonded particle assembly. The same surface was used for each wall of the joint. To be assigned a smooth-joint model, a contact must satisfy two conditions: (1) the two contacting balls must lie on opposite side of the plane containing the triangle; and (2) the projection of the contact location onto this plane must lie within the bounds of the triangle. The orientation of the smooth joint, defined by a dip angle and a dip direction, corresponds to the orientation of the triangle. The process is repeated for every triangle of the surface. The joint surface is hence modeled as a collection of smooth-joint contacts. The discontinuity was considered to be purely frictional (i.e. no bond is introduced) with a friction angle set to 20°. No dilation was introduced as macroscopic dilation (i.e. at the joint level) is expected to be an emergent property of the surface topology. The SJM parameters are given in Table 3. The output is a 140 mm × 140 mm synthetic rock joint sample (SRJ) whose morphology corresponds to the natural rock joint that is being analyzed.

Table 3. Smooth-joint model contact parameters.

Friction angle, $\phi_j(°)$	Dilation angle, $\psi_j(°)$	Radius multiplier, λ	Bond mode, M	Bond cohesion, c_j(MPa)
20	0	1	0	0
Bond tensile strength, σ_{cj} (MPa)	Bond friction angle, ϕ_{bj} (°)	Joint normal stiffness, k_{nj} (GPa/m)	Joint tangential stiffness, k_{tj} (GPa/m)	Large strain flag, Bl
0	0	50	12.5	1

In this paper, "synthetic rock joint" or SRJ will refer to the discrete element model of a rock joint. Its properties such as strength or stiffness will be macro-properties (i.e. computed at the scale of sample) and are denoted using an uppercase letter (e.g. Φ_{peak}, Ψ_{peak}). The SJM on the other hand refers to a local contact on the joint surface. Lowercase letters will be used for micro-properties (e.g. ϕ_j, k_{nj}). A full 3D view of the numerical sample can be seen in Fig. 3.

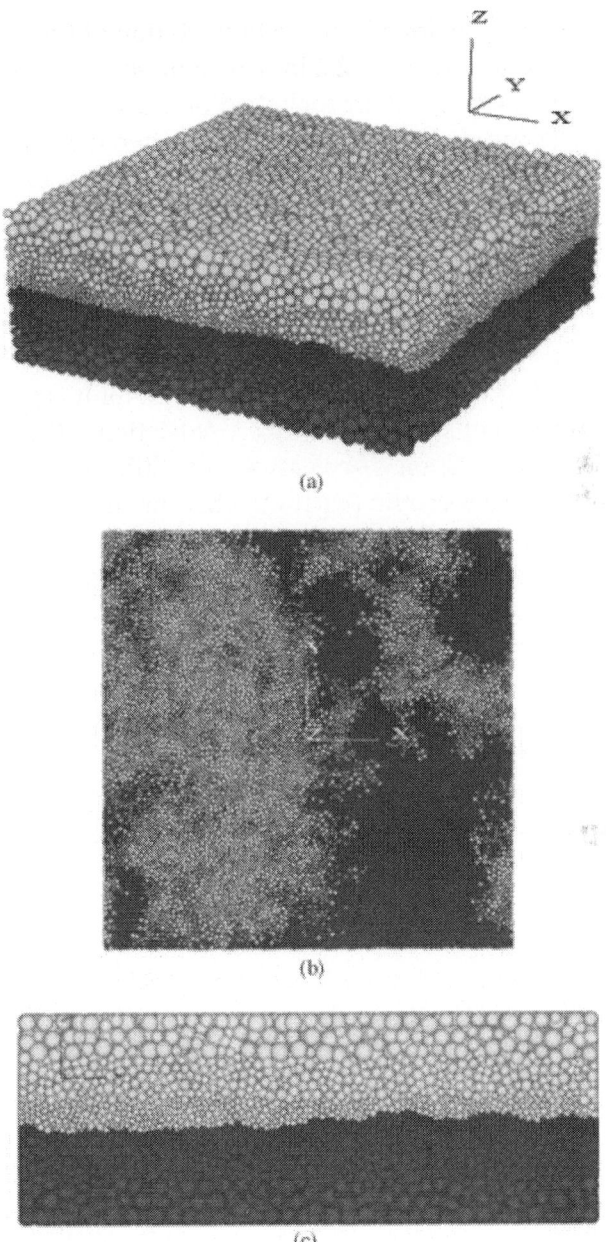

Figure 3. Visualization of the synthetic rock joint sample (upper wall in orange and lower wall in brown): (a) general 3D view; (b) horizontal cross-section through mean fracture plane; (c) vertical cross-section.

During the direct shear tests, specimens are firstly subjected to a compression along axisZ (Fig. 3) and then to a shearing along axis X at constant normal stress. During the compression stage, the normal load is applied to the upper wall of the specimen while displacements of the lower wall are restraint. The required load is applied in five incremental stages. For each stage, the incremental normal force is equally shared between the particles of the top layer of the specimen (i.e. between the particles whose centers are within one average diameter from the top of the specimen) and progressively applied in 100 time steps. The system is then dynamically set to equilibrium before proceeding to the next stage. During shearing, displacements along Y-axis are restrained whereas shear displacements along X-axis are applied to the lower wall. The sum of contact forces on the periphery of the upper wall is used to compute the average normal stress and shear stress on the interface whereas relative normal and tangential displacements are monitored, averaging particle displacements on the periphery of the lower wall (Z-displacements and X-displacements, respectively). Joint aperture is defined as the relative normal displacement. A particle is defined as belonging to the periphery if the distance from its center to the closest specimen boundary is lower than average diameter. Micro-cracks due to bond breakage, contact force distribution and stress–strain path are monitored during the shear tests. The direct shear tests are run in a large strain mode. As shear displacement increases, new contacts are created along the discontinuity. These contacts are assigned a smooth-joint contact model and the orientation of the smooth joint depends on the location of the contact. A special algorithm was developed to determine which triangle of the surface morphology is intersected by the newly created contact. As shearing occurs, the mirror surface associated with the upper wall does not match the lower surface. Each contact intersects two triangles with possibly different orientations. The orientation of each new contact could be associated in reality with any of the two surfaces or be a combination of the two surfaces. In this study, the assumption was made to consider only the surface morphology associated with the upper wall.

Mechanical behavior of the discrete interface

Numerical shear tests under constant normal stress were performed on the SRJ for three values of normal confinement. Normal stress values of 0.5 MPa, 1 MPa and 1.5 MPa were applied to the sample which correspond approximately to 0.35%, 0.7% and 1.05% of the intact rock UCS, respectively. Those low values of normal stress corresponding to the order of magnitude of normal stress practitioners usually have to be dealt with for slope stability problems. Fig. 4 shows the evolution of shear stress and normal displacement with shear displacement for one particular packing. It can be seen that the classical elasto-plastic response of rock joints is well captured, thus confirming a good agreement with typical behaviors that can be observed experimentally. The mobilized shear stress increases to a peak value as roughness is mobilized and then decreases due to asperity degradation. The peak value defines the shear strength of the SRJ (the higher the normal stress, the higher the shear strength). It can be noted that the peak is reached after 1.5 mm of tangential displacement which is on the upper limit of what is usually observed. The stiffness of the smooth joint was probably underestimated and this question will be discussed in Section 3.2. The peak shear strengths were 0.51 MPa, 0.9 MPa and 1.21 MPa. The friction calculated from the ratio of peak shear stress to applied normal stress was higher at lower normal stresses (1.01, 0.90 and 0.81 at 0.5 MPa, 1 MPa and 1.5 MPa, respectively), enhancing a nonlinear strength envelope.

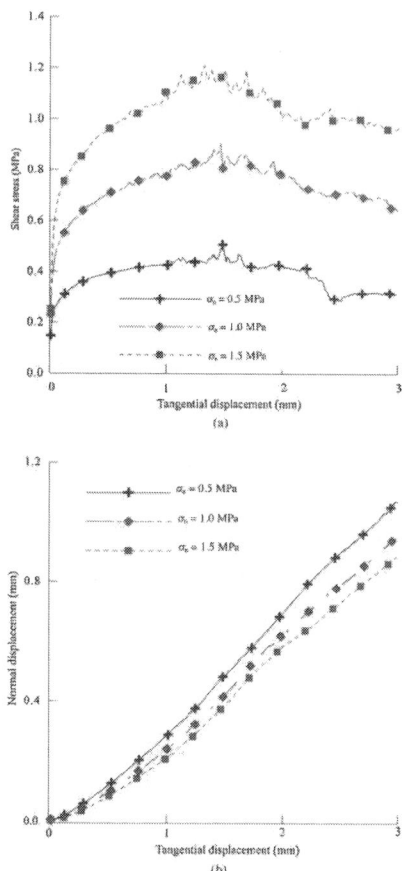

Figure 4. Stress and displacement curves of direct shear tests under constant normal stress (ranging from 0.5 MPa to 1.5 MPa) on a 140 mm × 140 mm surface. (a) Shear stress versus tangential displacement. (b) Normal displacement versus tangential displacement.

Normal displacement versus tangential displacement curves in Fig. 4 show that overall dilation of the rock joint is reduced as normal stress increases. As shearing takes place and roughness is mobilized, the dilation angle defined as $\delta u_n / \delta u_s$ increases to a maximum value Ψ_{peak} (peak dilation angle) at the peak of the shear stress.

Three different specimens were generated varying the packing of the particle assembly. The same shear tests were simulated on each specimen. Peak shear strengths τ_{peak} and peak dilation angles Ψ_{peak} are reported in Fig. 5. Peak shear strength shows a limited sensitivity to particle packing whereas significant variation of peak dilation angle is observed.

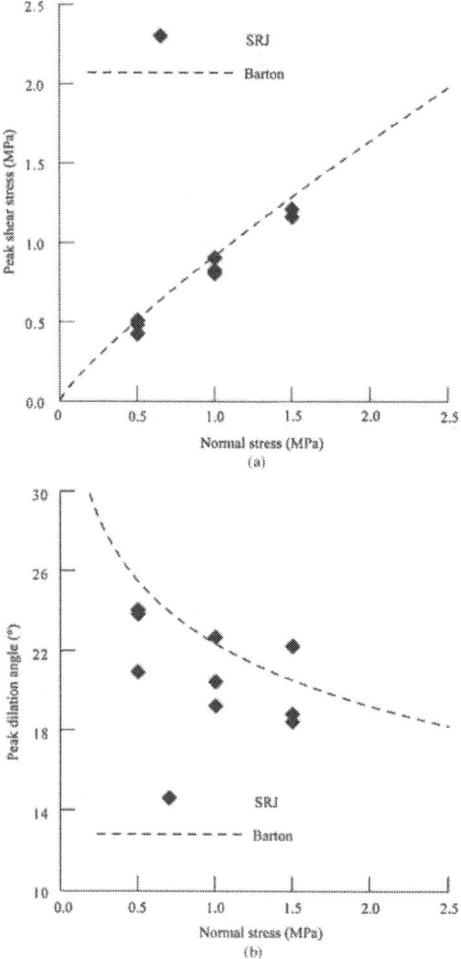

Figure 5. Comparison of peak shear strength and dilation angles of the synthetic rock joint (diamond) with their respective predicted values from Eqs. (3) and (4) (dash line).

The numerical shear tests performed under increasing normal stress define the strength envelope of the model from which a Barton failure criterion (Barton and Choubey, 1977) can be expressed. In

Barton's formulation, the shear strength is expressed as a function of the JRC, joint compressive strength (JCS) and base friction Φ_b:

$$\tau_p = \sigma_n \tan\left[JRC \log_{10}\left(\frac{JCS}{\sigma_n} + \Phi_b \right) \right]$$

(3)

where τ_p is the peak shear stress and σ_n the normal stress.

Barton's failure criterion was applied to predict the strength of the SRJ. Φ_b refers to the base friction of the joint which corresponds to the friction a perfectly smooth joint would have. The base friction of the SRJ corresponds to the friction angle of the SJM ($\phi_j = 20°$). The JCS corresponds to the UCS of the synthetic material, 143 MPa (Table 1), and the JRC of the triangulated surface was estimated to 10.4. In Barton's formulation, the dilation angle can be estimated using the following empirical relation:

$$\Psi_{peak} = JRC \log_{10}\left(\frac{JCS}{\sigma_n} \right)$$

(4)

The predictions of shear strength and peak dilation angle of the SRJ with Eqs. (3) and (4)were compared with the numerical results obtained from the simulations and can be seen in Fig. 5. The measured strength appears to be in very good agreement with the prediction obtained with a widely used relationship such as Barton's failure criterion. Fig. 5 shows some differences between the measured dilation angles and Barton's predictions. If for the range of normal stress applied in this study, the measured and predicted values are of the same order, the general trend in the decrease of dilation with normal stress is significantly different. An overestimation of the dilation can be expected at high normal stress. However, the SRJ seems to well capture the mechanical behavior of a natural rock joint.

PARAMETRIC STUDY

The SRJ, as described in Section 2, does not rely on any empirical scheme or any particular assumption on surface roughness. It is generated using a 3D measurement of the surface morphology and intact rock properties.

Such discrete model seems to well capture the effect of surface roughness on the mechanical behavior of rock joints. Results of shear test simulations show a very good agreement with Barton's prediction, based on JRC. This suggests that predictive estimations of shear strength should be possible combining 3D surface measurements with a smooth-joint contact model. A number of SRJ samples were generated varying the properties of the SJM and the particle friction angle to enhance the relation between some contact properties and the emergent macroscopic behavior. The same surface morphology (140 mm × 140 mm) was introduced and the same particle size distribution was used to generate the particle assembly representing the rock. Scale dependency and particle discretization will be discussed in Section 4. A SJM is defined through five parameters, i.e. friction, cohesion, dilation and stiffness (normal and tangential). Only purely frictional joint was considered at this stage. No local dilation (i.e. at the contact level) was introduced through the SJM as macroscopic dilation (i.e. at the joint level) is expected to be an emergent property of the surface's roughness. Effects of joint friction angle ϕ_j, joint stiffnesses k_{nj} and k_{tj}, and particle friction ϕ_p were analyzed.

Effect of joint friction angle

Four 140 mm × 140 mm joints were generated with a joint friction angle ϕ_j ranging from 15° to 30°. Direct shear tests were performed at a constant normal stress of 1.5 MPa. The evolution of shear stress and normal displacement with tangential displacement is shown in Fig. 6. As joint friction increases, peak shear strength and residual shear strength increase from 1.06 MPa to 1.60 MPa and from 0.82 MPa to 1.16 MPa, respectively. The residual shear strength can be characterized by relatively stable shear and normal stresses with degradation on joint surfaces still occurring (Gentier et al., 2000). Direct shear tests by Grasselli (2001) show that residual strength is reached slightly before 3 mm of tangential displacement. In this study, however, residual shear strength was defined as the shear stress after 3 mm of tangential displacement.

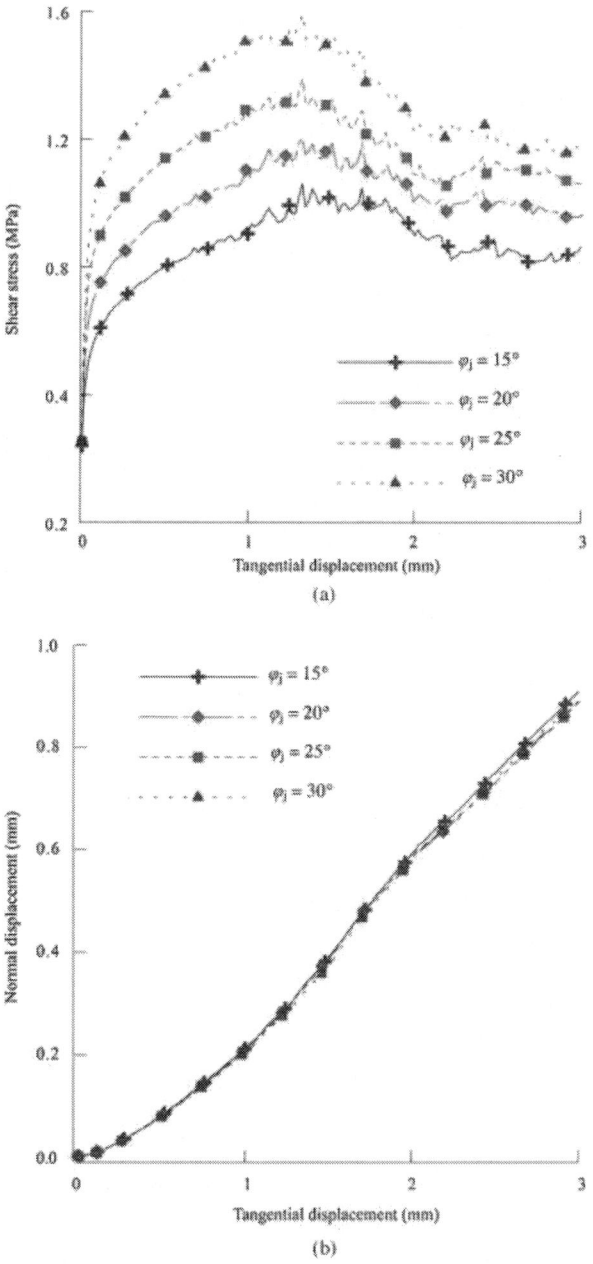

Figure 6. Stress and displacement curves of direct shear tests under constant normal stress (σ_n = 1.5 MPa) on a 140 mm × 140 mm surface with a smooth-joint friction angle ranging from 15° to 30°. (a) Shear stress versus tangential displacement. (b) Normal displacement versus tangential displacement.

As ϕ_j increases, the mechanical behavior becomes more brittle. As expected, dilation remains unchanged and emerges as independent of the smooth-joint friction angle. In comparison simulations by Park and Song (2009) exhibited a significant increase in dilation when friction coefficient increased from 0 to 0.3 (from 0° to 16.7°).

Mobilized peak friction angles Φ_{peak} (ratio between peak shear stress and normal stress) and mobilized residual friction angles Φ_{res} (ratio between residual shear stress and normal stress) can be seen in Fig. 7. Φ_{peak} varies from 35.4° to 46.8° whereas Φ_{res} varies from 28.8° to 37.8°. A very good linear relation can be drawn between the mobilized friction angles and the smooth-joint friction angle (coefficient of determination of 0.999 for Φ_{peak} and 0.987 for Φ_{res}). This result is consistent with the idealized decomposition of rock joint strength as the addition of a frictional component and an asperity component as suggested by Barton and Bandis (1982) and confirms that the smooth-joint friction angle ϕ_j should be calibrated according to base friction (i.e. friction angle of a planar surface).

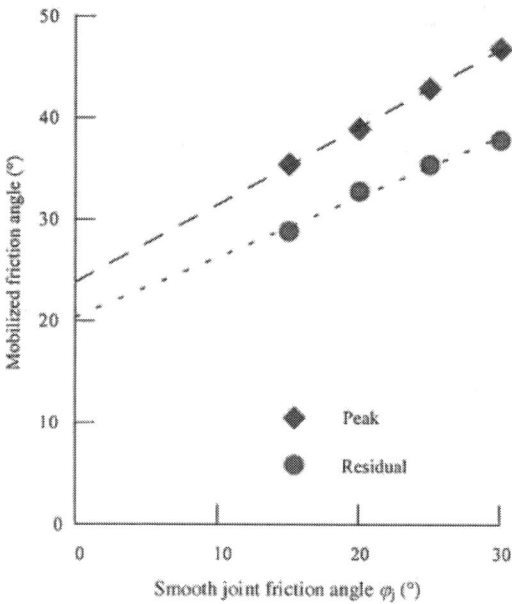

Figure 7. Peak and residual mobilized friction angle vs. smooth-joint friction angle ϕ_j at normal stress of 1.5 MPa. Diamonds and disks represent model values and dash lines best linear fit.

Effect of joint normal and tangential stiffness

To investigate the influence of smooth-joint normal and tangential stiffnesses on the macroscopic mechanical behavior, four sets of stiffnesses were used. The first three samples were generated varying the joint normal stiffness, i.e. 50 GPa/m, 250 GPa/m and 500 GPa/m corresponding to k_{nj}^0, $5k_{nj}^0$ and $10k_{nj}^0$, respectively. The tangential stiffness k_{tj}^0 was kept unchanged. For the last sample, a factor of 10 was applied to both k_{nj}^0 and k_{tj}^0 ($k_{nj} = 500$ GPa/m and $k_{tj} = 125$ GPa/m) which, by comparing with the case $10k_{nj}^0$ and k_{tj}^0, will provide information on the influence of the tangential stiffness. Stiffness values for each specimen are summarized in Table 4. The samples were submitted to a numerical shear test under a constant normal stress of 1.5 MPa. The evolution of shear stress and normal displacement upon shearing are shown in Fig. 8.

Table 4. Normal and tangential stiffnesses for smooth-joint contact model.

Sample	k_{nj} (GPa/m)	k_{tj} (GPa/m)
k_{nj}^0 and k_{tj}^0	50	12.5
$5k_{nj}^0$ and k_{tj}^0	250	12.5
$10k_{nj}^0$ and k_{tj}^0	500	12.5
$10k_{nj}^0$ and k_{tj}^0	500	125

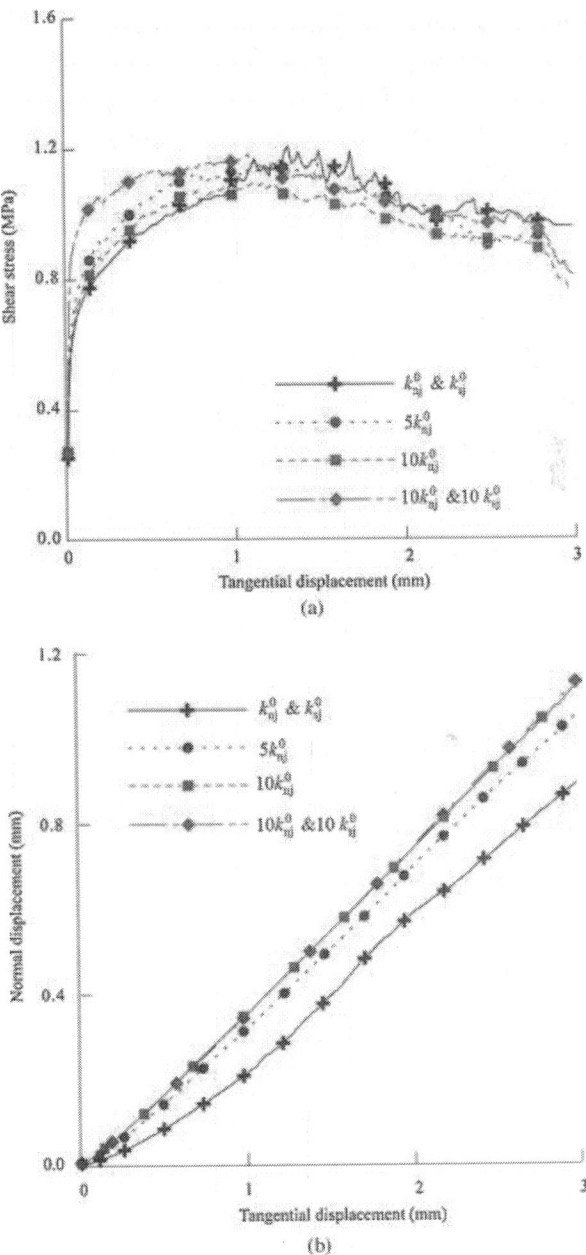

Figure 8. Stress and displacement curves of direct shear tests under constant normal stress ($\sigma_n = 1.5$ MPa) on a 140 mm × 140 mm surface with different smooth-joint normal and tangential stiffnesses. (a) Shear stress versus tangential displacement. (b) Normal displacement versus tangential displacement.

As k_{nj} increases, the overall tangential stiffness of the joint slightly increases. k_{nj} also seems to have a slight influence on the peak and residual shear strengths. The peak shear strengths were 1.21 MPa, 1.15 MPa and 1.10 MPa for k_{nj}^0, $5k_{nj}^0$ and $10k_{nj}^0$, respectively. The residual shear strength were 0.97 MPa, 0.82 MPa and 0.74 MPa for k_{nj}^0, $5k_{nj}^0$ and $10k_{nj}^0$, respectively. The relatively small variation in the peak and residual strength (9% and 14%, respectively) induced by a change of one order of magnitude in the normal stiffness can probably be attributed to a stress redistribution between the two walls of the joint. Fig. 9 shows the distribution of shear forces across the discontinuity after 1.4 mm of tangential displacement. As normal stiffness k_{nj} increases, the number of asperities interacting is reduced. The maximum contact shear force in the joint increases from 72.7 N to 227.8 N. As contact becomes stiffer, the true contact surface between the walls is slightly reduced hence generating higher local stresses on the asperities for the same external load. In addition, k_{nj} exhibits a significant influence on the dilation of the joint (see Fig. 8). If the peak dilation angle does not show any variation, the normal aperture of the joint (normal relative displacement) is significantly controlled by the normal stiffness. Final aperture (measured after 3 mm tangential displacement) rises from 0.89 mm to 1.12 mm.

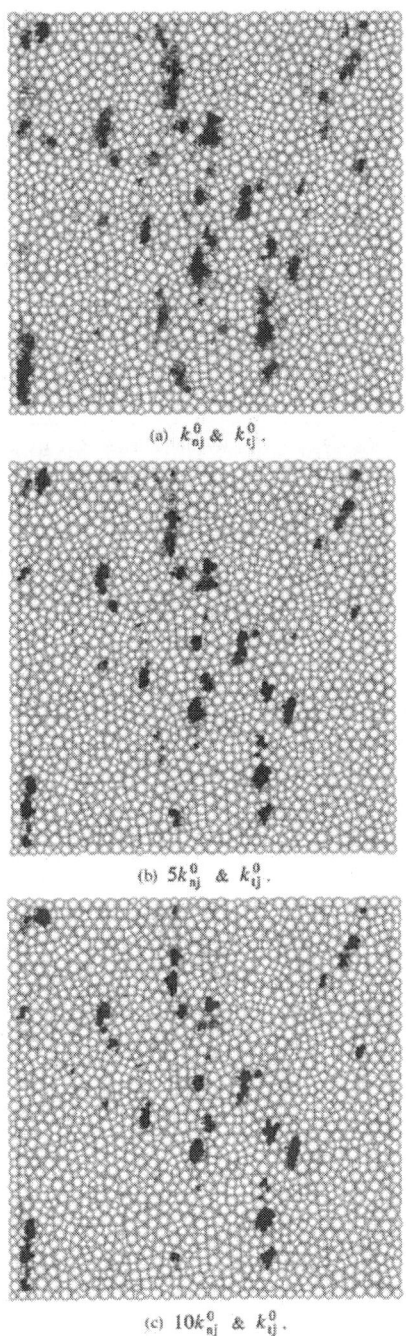

(a) k_{nj}^0 & k_{tj}^0.

(b) $5k_{nj}^0$ & k_{tj}^0.

(c) $10k_{nj}^0$ & k_{tj}^0.

Figure 9. Contact shear force distribution across the joint surface at a tangential displacement of 1.4 mm.

Fig. 8 shows that an increase of the SJM tangential stiffness increases significantly with the overall stiffness of the synthetic rock joint. No variation in the dilation can be observed. Both tests exhibit the same dilation angles and the final joint aperture is 1.12 mm for $10k_{nj}^0 - k_{ij}^0$ and 1.13 mm for $10k_{nj}^0 - k_{ij}^0$. A slight increase in the shear strength can be observed. Peak shear strength increases from 1.10 MPa to 1.19 MPa whereas residual shear strength increases from 0.82 MPa to 0.87 MPa. Interestingly, this variation is opposite to what was observed when increasing the SJM normal stiffness k_{nj}.

Effect of particle friction

Park and Song (2009) suggested that for planar surfaces particle–particle friction coefficient ($\mu = \tan \phi_p$) should be determined between 0.0 and 0.15 as significant increase in the shear strength is obtained and very little variation of the dilation angle is observed. In this study, direct shear tests on rough synthetic rock joints were performed under a constant normal stress of 1.5 MPa. Particle friction angles of 0°, 5° and 10° were used, corresponding to friction coefficients of 0.0, 0.087 and 0.176, respectively. For each particle friction angle, the micro-properties were calibrated to generate particle assemblies exhibiting the same mechanical behavior (deformation and strength) on unconfined compression test. The evolutions of the shear stress and normal displacement upon shearing are shown in Fig. 10.

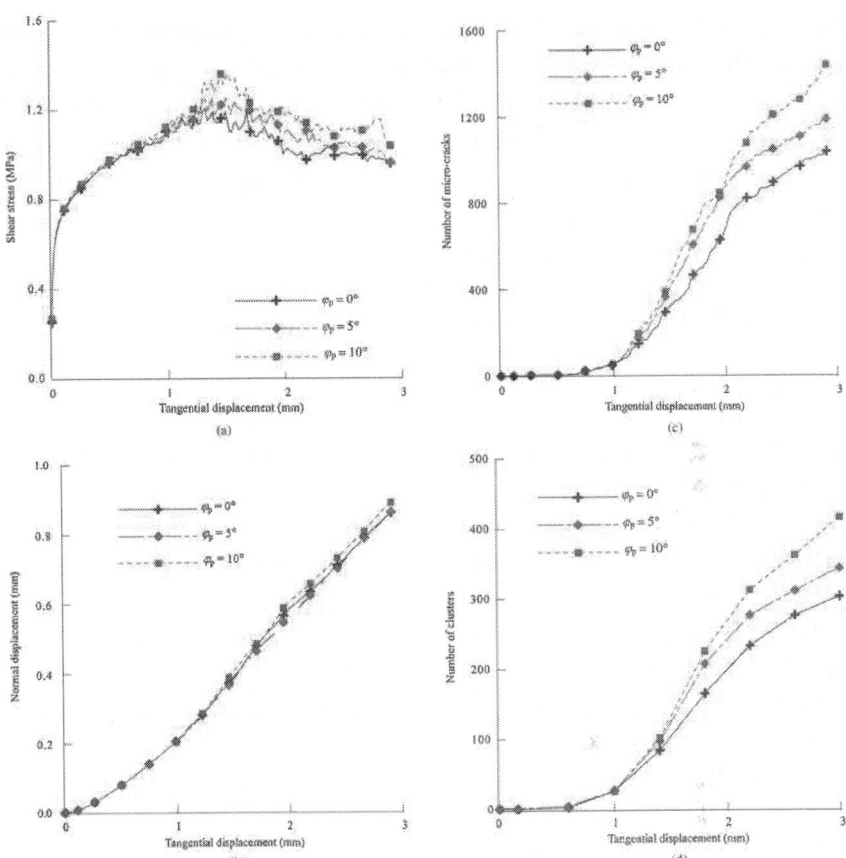

Figure 10. Stress and displacement curves of direct shear tests under constant normal stress ($\sigma_n = 1.5$ MPa) on a 140 mm × 140 mm surface for different particle friction angles (σ_p ranging from 0° to 10°). (a) Shear stress versus tangential displacement. (b) Normal displacement versus tangential displacement. (c) Number of micro-cracks versus tangential displacement. (d) Number of clusters in the gouge vs. tangential displacement.

Peak and residual shear strengths increase with particle friction (ϕ_j kept constant). No significant difference materializes on the shear stress curves before a tangential displacement of 1 mm. After 1 mm, micro-cracking (i.e. bond breakage) in the particle assembly becomes significant as can be seen in Fig. 10. The increase of micro-cracks leads to the formation of a gouge (materialized as single particles or clusters of several particles) between the two walls of the joint. Differences in the number of clusters forming the gouge appear after a tangential displacement of 1 mm (see Fig. 10d). It can be noted that significant degradation occurs long after the peak shear stress. The rate of degradation however is maximal at or immediately after the peak. With the formation of a gouge, forces across the interface are transmitted not only through smooth-joint contacts but also through clusters of particles. Shear strength of filled joints is highly influenced by the strength of the infill material. As the particle friction angle is increased, the strength of the newly formed gouge is increased. The dilation shows little variation for the range of particle friction angles studied here. However significant increase can be expected for higher values of ϕ_p as stated in Park and Song (2009).

LARGE-SCALE DISCONTINUITIES

Bandis et al. (1981) identified two contributors to rock joint strength: a basic frictional component (base friction) and a roughness component (Fig. 11).

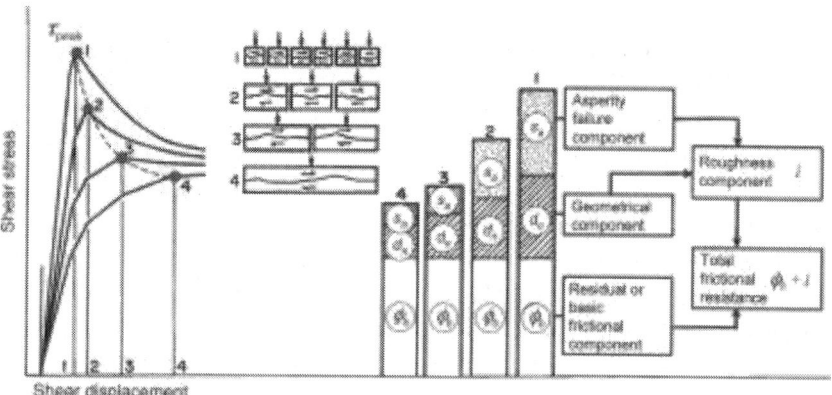

Figure 11. Scale effects in the shear strength components of non-planar defects (after Bandis et al. (1981)).

Geometry (or morphology) of the discontinuity (shape of the asperities) and asperity failure (the strength of the asperities) are the basis of the roughness component. When base friction appears to be scale-independent and can be estimated on laboratory-scale experiments, the roughness component is highly scale-dependent. Roughness decreases as scale increases (Bandis et al., 1981). Numerous studies were carried out trying to quantify the scale dependence of joint strength from which empirical relations were proposed (Barton and Bandis, 1982):

$$JRC_n = JRC_0 \left(\frac{L_n}{L_0} \right)^{-0.02\,JRC_0}$$

(5)

$$JCS_n = JCS_0 \left(\frac{L_n}{L_0} \right)^{-0.03\,JRC_0}$$

(6)

Because of the scale dependency observed in the mechanical behavior of discontinuities, their properties should be assessed at the relevant scale. In a rock mass, the scale of the discontinuities ranges from meters to hundreds of meters (and more). Laboratory methods, where scale is usually restricted to meter and below, cannot be directly extended for field estimates.

Scale dependency of SRJ behavior

In this study, scale dependency of the model was investigated performing numerical shear tests on samples of various sizes. Two smaller scales were tested, 70 mm × 70 mm and 46.7 mm × 46.7 mm, splitting the initial surface into respectively four and nine sub-surfaces which were imported into a bonded-particle assembly. Same micro-properties and particle size distribution were used to represent the intact rock. Direct shear tests under a constant normal stress of 1.5 MPa were performed on each of the four + nine newly created synthetic rock joints. Fig. 12 shows peak shear stress (mean value and range of distribution) and peak dilation angle (mean value and range of distribution) versus sample size for all the tests.

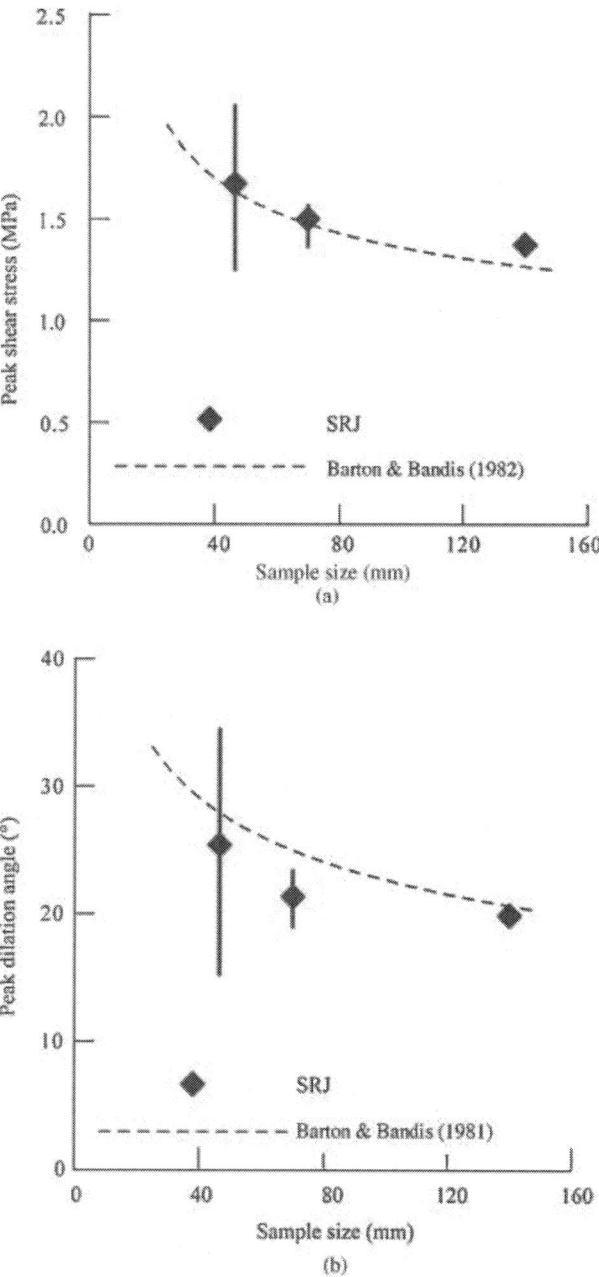

Figure 12. Variation of peak shear stress and peak dilation angle with joint size (σ_n = 1.5 MPa): mean value (diamond) and variability (plain line). Comparison with combined empirical relations (Eqs. (3), (4), (5) and (6)) (afterBarton and Choubey, 1977 and Barton and Bandis, 1982).

Peak shear strength and peak dilation decrease significantly for a sample size increasing from 46.7 mm to 140 mm. Mean peak shear stress dropped from 1.67 MPa to 1.37 MPa, corresponding to a 17.8% decrease. Peak dilation angle dropped from 25.4° to 19.9°, corresponding to a 21.8% decrease. Variability in peak and dilation angles is reduced as sample size increases. Combining the empirical relations Eqs. (3), (4), (5) and (6), predictions on the scale dependency of peak friction angle and peak dilation angle are shown in Fig. 12. The SRJ exhibits a scale dependency of its mechanical properties in good agreement with predictions based on empirical relations. However, peak dilation angles of the SRJ appear slightly lower than those predicted with Barton and Bandis' relations.

Effect of particle size

With the development of laser measurement systems and 3D photogrammetry techniques, practitioners can now have access to topological descriptions of large discontinuities (meter and above). These large-scale morphologies could be used to generate large synthetic rock joints. Estimates of their shear strengths could hence be derived without the need for any empirical relations. However with the current computer limitations, a limited number of particles can be used, and testing these large-scale discontinuities with DEM would require the use of larger particles. As particles become larger than the smallest asperities, surface roughness can be artificially reduced. Direct shear tests under constant normal stress (1.5 MPa) were performed on 70 mm × 70 mm samples using different particle size distributions, with an average diameter in the vicinity of the interface ranging from 1.29 mm to 2.56 mm. The shape of the particle size distribution (ratio between minimum and maximum radius of 1.66) and micro-properties were kept unchanged.

Fig. 13 shows the peak shear stress and the peak dilation angle versus the average particle diameter. Dilation is reduced as particle size increases which tends to confirm that the surface roughness of the joint is reduced as particles becomes larger. The horizontal spacing between two grid points of the surface is 1.4 mm whereas the lowest average diameter used for these simulations is 1.29 mm, resulting in approximately one particle per interval. No asymptotic value appears when decreasing the particle diameter, thus suggesting that a minimum of two (or more) particles per interval should be introduced to capture the full roughness of the surface. Interestingly, shear strength exhibits no such trend. In comparison, shear test simulations by Park and Song (2009)show no conclusive effect on peak friction angle and peak dilation angle. A full understanding of the effect of particle size with the SJM requires additional analyses.

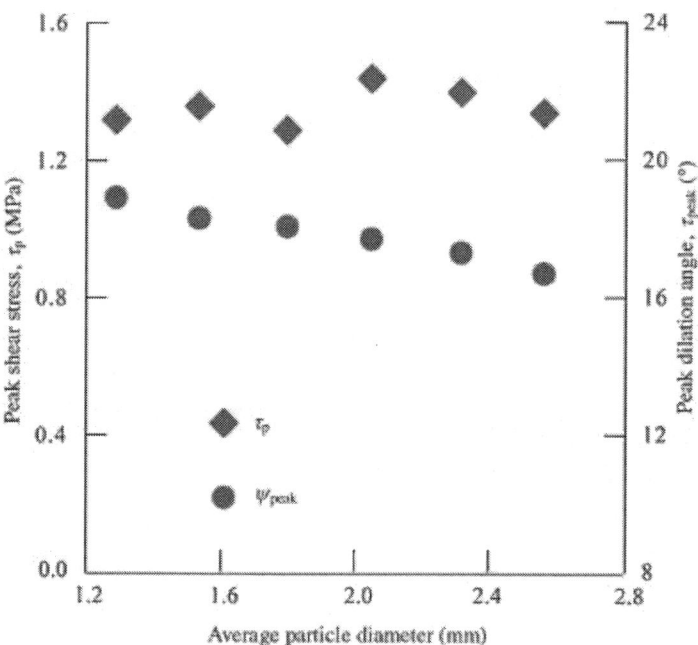

Figure 13. Evolution of peak shear stress τ_p and peak dilation angle ψ_{peak} with average particle diameter. 70 mm × 70 mm sample and σ_n = 1.5 MPa.

Significance for large discontinuities

The surface shape or topology of a discontinuity is seen at the micro-scale level as a series of asperities and results from two different components of surface texture, roughness and waviness (Belem et al., 2009). The roughness component is termed "secondary" or second order surface roughness, and the waviness component is termed "primary" or first order surface roughness. Both orders of asperities have to be taken into consideration when considering joint roughness and thus joint strength (Plesha, 1987,Yang et al., 2001 and Haneberg et al., 2007). The second order asperities exhibit high angles and narrow base lengths (or wave length) in opposition to the first order asperities that have lower angles and longer base lengths. The behavior of rock joints is controlled primarily by the second order asperities during small displacements, and first order asperities govern the shearing behavior for large displacements. Barton and Choubey (1977) first stated that at low normal stress levels, the second order asperity controls the shearing process. With increasing normal stress, the second order asperity is sheared off and the first order asperity takes over as the controlling factor. Fardin et al. (2001)suggested that a resolution of 0.2 mm in the roughness measurement was required to correctly capture the second order asperities whereas a resolution of 20 mm seems sufficient to capture the first order asperities. Yang et al. (2001) obtained similar conclusions using analytical decompositions.

With the current computer capacities, capturing the effect of the second order asperities for large joints is currently not achievable. However, the first order asperities could be accurately described for joint surfaces of 1 m² and above. Strength characterizations would then be restrained to situations where primary asperities are the controlling factor.

CONCLUSIONS

In this paper, a new DEM representation of rock joints was presented. Numerical joints were generated combining a real 3D surface morphology and the smooth-joint contact model. Particles lying on opposite side of the joint surface were assigned a smooth-joint contact. At macro-scale level, the behavior of a natural joint is a combination of surface roughness, intact rock properties and frictional contact behavior. The behavior of the synthetic rock joint is an emergent property of the surface morphology, micro-properties of the particle assembly and micro-properties of the smooth-joint contact model. The surface morphology introduced was measured by Grasselli (2001) using a laser scanner. Micro-properties of the particle assembly were calibrated on the basis of measured intact rock properties (UCS, Young's modulus).

Direct shear tests under constant normal stress were simulated and the mechanical response of the discrete model was analyzed. The shear behavior was compared to the expected behavior of a joint with the same morphology, the latter being assessed with conventional JRC based estimation methods. A relatively good agreement could be established. The effect of rou

ACKNOWLEDGEMENTS

The authors would like to thank Itasca Consulting Group Inc. for their technical advice and Dr. Giovanni Grasselli for providing the natural morphology of a discontinuity in granite whose work was performed with funding provided by the Swiss Federal Office for Water and Geology.

REFERENCES

1. Bandis S, Lumsden A, Barton N. Experimental study of scale effects on the shear behavior of rock joints. International Journal of Rock Mechanics and Mining Sciences & Geomechanics Abstracts 1981;18(1):1–21.
2. Barton N, Choubey V. The shear strength of rock joints in theory and practice. Rock Mechanics 1977;10(1–2):1–54. Barton N. Review of a new shear strength criterion for rock joints. Engineering Geology 1973;7(4):287–332.
3. Barton N, Bandis SC. Effect of block size on the shear behavior ofjointed rocks. In: The 23rd US Symposium on Rock Mechanics. Berkeley: American Rock Mechanics Association; 1982. p. 732–60.
4. Belem T, Souley M, Homand F. Method for quantification of wear of sheared joint walls based on surface morphology. Rock Mechanics and Rock Engineering 2009;42(6):883–910.

5. Buzzi O, Hans J, Boulon M, Deleruyelle F, Besnus F. Hydromechanical study of rock-mortar interfaces. Physics and Chemistry of the Earth 2008;32(8–14): 820–31.

6. Carr JR, Warriner JB. Relationship between the fractal dimension and joint roughness coefficient. Bulletin of the Association of Engineering Geologists 1989;26(2):253–63.

7. Cundall P. Numerical experiments on rough joints in shear using a bonded particle model. In: Lehner FK, Urai JL, editors. Aspects of Tectonic Faulting. Berlin: Springer-Verlag; 2000. p. 1–9.

8. Deisman N, Ivars MD, Darcel C, Chalaturnyk RJ. Empirical and numerical approaches for geomechanical characterization of coal seam reservoirs. International Journal of Coal Geology 2010;82(3–4):204–12.

9. Dershowitz W. Interpretation and synthesis of discrete fracture orientation, size, shape, spatial structure and hydrologic data by forward modeling. In: Cook NGW, Goodman RE, Myer LR, Tsang CF, editors. Fractured and Jointed Rock Masses. Netherlands: A.A. Balkema; 1995. p. 579–86.

10. Fardin N, Stephansson O, Jing L. The scale dependence of rock joint surface roughness. International Journal of Rock Mechanics & Mining Sciences 2001;38(5):659–69.

11. Fardin N, Feng Q, Stephansson O. Application of a new in situ 3D laser scanner to study the scale effect on the rock joint surface roughness. International Journal of Rock Mechanics & Mining Sciences 2004;41(2):329–35.

12. Gentier S, Riss J, Archambault G, Flamand R, Hopkins D. Influence of fracture geometry on shear behavior. International Journal of Rock Mechanics & Mining Sciences 2000;37(1–2):161–74.

13. Giacomini A, Buzzi O, Krabbenhoft K. Modelling the asperity degradation of shear rock joint using FEM. In: The 8th World Congress on Computational Mechanics (WCCM8) and the 5th European Congress on Computational Methods in Applied Sciences and Engineering (ECCOMAS 2008); 2008.

14. Grasselli G. Shear strength of rock joints based on quantified surface description. PhD thesis. Switzerland: Ecole Polytechnique Federale de Lausanne; 2001.

15. Grasselli G, Wirth J, Egger P. Quantitative three-dimensional description of a rough surface and parameter evolution with shearing.

International Journal of Rock Mechanics & Mining Sciences 2002;39(6):789–800.

16. Grasselli G, Egger P. Constitutive law for the shear strength of rock joints based on three-dimensional surface parameters. International Journal of Rock Mechanics & Mining Sciences 2003;40(1):25–40.

17. Haneberg WC. Directional roughness profiles from three-dimensional photogrammetric or laser scanner cloud points. In: Rock Mechanics: Meeting Society's Challenges and Demands, Proceedings of the 1st Canada-US Rock Mechanics Symposium. Vancouver, Canada: Taylor & Francis; 2007. p. 101–6.

18. Hans J, Boulon M. A new device for investigating the hydromecanical properties of rock joints. International Journal for Numerical and Analytical Methods in Geomechanics 2003;27(6):513–48.

19. Hutson R, Dowding C. Joint asperity degradation during cyclic shear. International Journal of Rock Mechanics and Mining Sciences & Geomechanics Abstracts 1990;27(2):109–19.

20. Itasca Consulting Group Inc. . Particle flow code in 3 dimensions version 4.0: User manual. Minneapolis, MN: Itasca Consulting Group Inc; 2008.

21. KaramiA, Stead D.Asperity degradation and damage in the direct shear test: a hybrid FEM/DEM approach. Rock Mechanics and Rock Engineering 2008;42(2):229–66.

22. Kulatilake P, Malama B, Wang J. Physical and particle flow modeling of jointed rock block behavior under uniaxial loading. International Journal of Rock Mechanics & Mining Sciences 2001;38(5):641–57.

23. Lambert C, Darve F, Nicot F. Rock slope stability from microscale to macroscale level. In: Pietruszczak P, editor. Numerical Models in Geomechanics. Ottawa: A.A. Balkema; 2004. p. 115–20.

24. Lambert C. Variability and uncertainty on rock mass strength via a synthetic rock mass approach. In: 1st Southern Hemisphere International Rock Mechanics Symposium. Perth, Australia: Australian Centre for Geomechanics; 2008. p. 355–66.

25. Lambert C, Coll C. Santiago A DEM approach to rock joint strength estimates Santiago. In: Rock Slope Stability in Open Pit Mining and Civil Engineering; 2009.

26. Lambert C, Buzzi O, Giacomini A. Influence of calcium leaching on the mechanical behavior of a rock mortar interface: a DEM analysis. Computers and Geotechnics 2010;37(3):258–66.

27. Lee H, Park Y, Cho T, You K. Influence of asperity degradation on the mechanical behavior of rough rock joints under cyclic shear loading. International Journal of Rock Mechanics & Mining Sciences 2001;38(7):967–80.

28. Misra A. Effect of asperity damage on shear behavior of single fracture. Engineering Fracture Mechanics 2002;69(17):1997–2014.

29. Park JW, Song JJ. Numerical simulation of a direct shear test on a rock joint using a bonded-particle model. International Journal of Rock Mechanics & Mining Sciences 2009;46(8):1315–28.

30. Patton FD. Multiple modes of shear failure in rock. In: Proc. 1st Int. Cong. Rock Mech. Lisbon: International Society for Rock Mechanics; 1966. p. 509–13.

31. Pierce M, Cundall P, Potyondy D, Ivars MD. A synthetic rock mass model for jointed rock. In: Rock Mechanics: Meeting Society's Challenges and Demands, Proceedings of the 1st Canada-US Rock Mechanics Symposium. Vancouver, Canada: Taylor & Francis; 2007. p. 341–9.

32. Pine R, Owen D, Coggan J, Rance J. A new discrete fracture modelling approach for rock masses. Geotechnique 2007;57(9):757–66.

33. Plesha M. Constitutive models for rock discontinuities with dilatancy and surface degradation. International Journal for Numerical and Analytical Methods in Geomechanics 1987;11(4):345–62.

34. Potyondy DO, Cundall PA. A bonded-particle model for rock. International Journal of Rock Mechanics & Mining Sciences 2004;41(8):1329–64.

35. Rogers S, Kennard D, Dershowitz W, van As A. Characterising the in situ fragmentation of a fractured rock mass using a discrete fracture network approach. In: Rock Mechanics: Meeting Society's Challenges and Demands, Proceedings of the 1st Canada-US Rock Mechanics Symposium. Vancouver, Canada: Taylor & Francis; 2007. p. 137–43.

36. Saeb S, Amadei B. Modelling rock joints under shear and normal loading. International Journal of Rock Mechanics & Mining Sciences 1992;29(3):267–78.

37. Tse R, Cruden C. Estimating joint roughness coefficients. International Journal of Rock Mechanics and Mining Sciences & Geomechanics Abstracts 1979;16(5):303–7.

38. Yang ZY, Di CC, Yen KC. The effect of asperity order on the roughness of rock joints. International Journal of Rock Mechanics & Mining Sciences 2001;38(5):745–52.

39. Yu X, Vayssade B. Joint profiles and their roughness parameters. International Journal of Rock Mechanics and Mining Sciences & Geomechanics Abstracts 1991;28(4):333–6.

40. Zhao Z, Jing L, Neretnieks I. Particle mechanics model for the effects of shear on solute retardation coefficient in rock fractures. International Journal of Rock Mechanics & Mining Sciences 2012;52:92–102.

41. Zhao Z. Gouge particle evolution in a rock fracture undergoing shear: a microscopic DEM study. Rock Mechanics and Rock Engineering 2013;46(6):1461–79.

CITATION

C. Lambert, C. Coll, Discrete modeling of rock joints with a smooth-joint contact model, Journal of Rock Mechanics and Geotechnical Engineering, Volume 6, Issue 1, February 2014, Pages 1-12, ISSN 1674-7755, http://dx.doi.org/10.1016/j.jrmge.2013.12.003.

Index